Foundations of Molecular Structure Determination

The front cover shows the three-dimensional structure and hand-drawn electron density map of phenoxymethylpenicillin obtained from x-ray diffraction experiments by Dorothy Hodgkin (kindly provided by Dr Adrian Whitwood).

Foundations of Molecular Structure Determination

SECOND EDITION

Simon Duckett

Bruce Gilbert

Martin Cockett

OXFORD

UNIVERSITY PRESS

OXFORD
UNIVERSITY PRESS

Great Clarendon Street, Oxford, OX2 6DP,
United Kingdom

Oxford University Press is a department of the University of Oxford.
It furthers the University's objective of excellence in research, scholarship,
and education by publishing worldwide. Oxford is a registered trade mark of
Oxford University Press in the UK and in certain other countries

First edition 2000

Impression: 1

Published in the United States of America by Oxford University Press
198 Madison Avenue, New York, NY 10016, United States of America

British Library Cataloguing in Publication Data

Data available

ISBN 978-0-19-968944-6

Printed in Great Britain by Ashford Colour Press Ltd, Gosport, Hampshire

Preface to the First Edition

This book is written primarily for those studying first-year University courses in Chemistry and for those preparing to do so.

It is designed to reflect significant recent advances in the use of spectroscopic and diffraction methods, not only for obtaining an analysis of elements and groups present in a molecule but also for establishing the arrangement of the constituent atoms. These developments have had a profound effect by increasing scientific knowledge in the fields of chemistry and molecular biology, reflected in the elucidation of the structure and function of a wide range of compounds including drugs, proteins and enzymes, and nucleic acids.

It is important that such work and its appreciation should find its place in the curriculum — as a means of educating chemistry students about essential principles and wide-ranging applications and to show how problem-solving skills are developed and applied in industrial and research environments. We also hope to convey the enjoyment and satisfaction of successful spectrum analysis.

We have included mass spectrometry and X-ray diffraction, along with traditional spectroscopic techniques; the former is the method *par excellence* for molecular mass and formulae determination and the latter provides details of molecular structure, providing information complementary to IR, NMR and UV-visible spectroscopy. We introduce the essential physical principles of each method, many examples of spectral analysis, and some problems; further reading and practice are strongly encouraged.

SI units have been adopted, with IUPAC nomenclature; trivial names accompany the recommended names in parentheses. Accurate mass data are taken from *Mass and Abundance Tables for Use in Mass Spectrometry* by J. H. Beynon and A. E. Williams, Elsevier, Amsterdam, 1963, and fragmentation patterns from *Compilation of Mass Spectral Data* by A. Cornu and R. Massot, Heyden, London, 1966.

We thank especially the following for assistance in recording spectra: Kin MyaMya, Anthony Crawshaw, ZygmuntDerewenda, Guy Dodson, Chris Hall, Reuben Girling, Rod Hubbard, Robert Liddington, and Ted Parton. We acknowledge permission to use Fig. 6.19 (The Royal Society). We are grateful for particularly relevant advice from schoolteachers including David Bevan, Michael Cane, Peter Gradwell, Geoff Liptrot, Bill Pickering, and George Walker. Finally, our special thanks go to David Waddington and Barry Thomas for enthusiasm and encouragement, and to Sue Street and Adrian Whitwood for assistance in producing the manuscript.

York
1999

S. B. D. and B. C. G.

Preface to the Second Edition

In preparing this second edition of our primer, we have been very conscious of our commitment to the initial aim 'to reflect recent advances in the use of spectroscopic and diffraction methods, not only for obtaining an analysis of the elements and groups present in a molecule but also for establishing the arrangement of constituent atoms'. This, again, is our major goal, with emphasis placed on the ways in which these approaches provide enhanced understanding and knowledge about molecular structure; and also, of course, how their application provides opportunities (especially for year-one university students) to learn and develop skills in structural analysis. As before, we have chosen to include chapters on mass spectrometry and the diffraction techniques (X-ray, neutron, and electron), as well as the spectroscopic methods of IR, UV, and NMR, to reflect the crucial information they can bring to structural analysis (e.g. on molecular weights, on empirical and structural formulae, and on 3-dimensional structural relationships).

In revising the text and content, we have aimed to provide many opportunities for practice: worked examples; encouragement for self-help; exercises with answers provided; and on-line multi-choice questions. Our policy has been to encourage practice in basic manipulation and understanding of information, before progressing to more advanced examples.

New sections have been included on Raman spectroscopy, alongside infra-red and microwave spectroscopy, and both neutron and electron diffraction, alongside X-ray diffraction. We have increased our coverage of enhancements in methodology and analysis, for example those resulting from advances in computer-based technology. In particular, we have extended our treatment of infra-red and Raman spectroscopy, placing more emphasis on spectroscopic principles involving, for instance, energy levels, populations, selection rules and transitions, and geometric parameters (e.g. bond lengths), as well as those used in structural recognition. We have also placed greater focus on the effects that molecular symmetry, size, and phase play in determining the kinds of information available and how specific methods can be employed to greatest effect.

The range of applications of these techniques has been extended, not least into biological and medical systems (for example in the development of magnetic resonance imaging, MRI). It has also been our intention to highlight the types of changes in instrumentation, to cover, for example, the demand for less expensive spectrometers to operate on a bench-top, as well as those 'top-of-the range' spectrometers for very high sensitivity or even those for operation in remote and hostile environments (e.g. on other planets). But the main emphasis remains on the student readers and their ability to understand and interpret the signals they get from the instruments. Our underpinning philosophy is that tackling problems takes readers to the core of the methods employed, and is likely to provide a stimulus for more advanced study, understanding, and application.

We thank, especially, those who have assisted us by providing helpful comments, or spectra, or both (e.g. via access to their papers in the primary literature); thus our gratitude is extended most warmly to our colleagues Ian Fairlamb, Brendan Keeley,

Jason Lynam, and Peter O'Brien, as well as John Moore and Derek Wann for helpful discussions. We are also indebted substantially to Adrian Whitwood for providing helpful comments, access to papers and computer files, and for computational and on-line assistance. We acknowledge most warmly the care and expertise of Lyndsay Muschamp and Katie Stott, for their invaluable assistance with the production of the manuscripts; and to the staff at OUP, most notably Alice Roberts, for their expertise and patient support.

York   S. B. D., B. C. G, and M. C. R. C
2015

Table of Contents

Preface to the Second Edition   vi

Chapter 1   Overview, energy levels, and the electromagnetic spectrum   1

1.1   Introduction   1

1.2   Energy levels, transitions between them, and the electromagnetic spectrum   2

Chapter 2   Rotational and vibrational spectroscopy   6

2.1   Introduction   6

2.2   Rotational spectroscopy   6

2.3   Rotational energy levels   7

2.4   Pure rotational microwave and millimetre wave spectroscopy of diatomic molecules   9

2.5   Microwave and millimetre wave spectroscopy of linear triatomic molecules   13

2.6   Rotational spectroscopy of non-linear polyatomic molecules   16

2.7   Experimental methods in microwave and millimetre wave spectroscopy   18

2.8   Rotational Raman spectroscopy   19

2.9   Experimental methods in rotational Raman spectroscopy   22

2.10   Vibrational spectroscopy   23

2.11   Infrared spectroscopy   24

2.12   Vibrational Raman spectroscopy   28

2.13   Vibration–rotation spectroscopy   30

2.14   Group vibrations, chemical characterization, and analysis   38

2.15   Examples of infrared spectra of organic molecules   44

2.16   Carbonyl group modes in inorganic metal complexes   47

2.17   Summary   48

2.18   Exercises   49

2.19   Further reading   51

Chapter 3   Electronic (ultraviolet–visible) absorption spectroscopy   52

3.1   Introduction   52

3.2   Electronic energy changes   53

3.3	Electronic absorption spectroscopy of organic molecules	53
3.4	The relationship of λ_{max} and ε_{max} to structure	57
3.5	Some applications of UV and visible absorption spectroscopy	62
3.6	Summary	66
3.7	Exercises	66
3.8	Further reading	67

Chapter 4 Nuclear magnetic resonance spectroscopy — 68

4.1	Introduction	68
4.2	The NMR experiment	68
4.3	^{1}H NMR spectra of organic molecules	71
4.4	Examples of spectra showing spin–spin splittings	80
4.5	Other structural information from ^{1}H NMR studies	85
4.6	NMR from other nuclei	88
4.7	Pulsed NMR spectrometers	89
4.8	2-Dimensional NMR methods	95
4.9	Magnetic resonance imaging (MRI)	98
4.10	Summary	99
4.11	Exercises	99
4.12	Further reading	101

Chapter 5 Mass spectrometry — 102

5.1	Introduction	102
5.2	The mass spectrometry experiment	102
5.3	Measuring relative molecular and atomic masses	105
5.4	Mass spectrometry of molecules: a detailed example	106
5.5	Analysis of mass spectra	107
5.6	Applications of mass spectrometry	119
5.7	Summary	126
5.8	Exercises	127
5.9	Further reading	129

Chapter 6 X-ray diffraction and related methods — 130

6.1	Introduction	130
6.2	Introduction to the X-ray diffraction method	130
6.3	Crystallography	135
6.4	Determination of structure	138
6.5	Structural determination for molecules	143
6.6	Neutron diffraction	147

6.7	Electron diffraction—method and structure determination	147
6.8	Summary	150
6.9	Exercises	151
6.10	Further reading	151
Glossary		153
Index		159

Overview, energy levels, and the electromagnetic spectrum

1.1 Introduction

Molecules come in a bewildering variety of shapes and sizes, and with a huge diversity of structural complexity. From the simplest homonuclear diatomic molecules such as H_2, to macromolecules such as haemoglobin containing many thousands of constituent atoms, the complexity of any molecule derives from the large number of potential constituent nuclei, the types of bonds which hold the nuclei together, and the degree to which the atoms arrange themselves into symmetric structures. Whilst the universe is seemingly bent on preferring to favour increasing disorder as the natural order of things, nature has a habit of creating molecules of considerable structural order and symmetry. Indeed, the extent to which any given molecule can be regarded as either unsymmetrical and disordered in its form or symmetrical, ordered and, yes, beautiful, is usually directly correlated with the efficiency with which it carries out its chemical functions.

Molecular bonding is absolutely fundamental to all aspects of chemistry, and it is the nature of the interactions that exist between nuclei, both weak and strong, which defines the types of structures that can be constructed from the atomic building blocks. Our understanding of chemical structure derives to a great extent from the theory of the chemical bond and, in particular, from the foundations laid by the development of quantum theory and quantum mechanics in the early part of the 20th century. However, it also depends critically on our ability to determine molecular structures through experiment.

This book presents a number of different experimental strategies to determine the geometrical arrangement of atoms in space that make up a particular molecule. The choice of which technique to use depends on a number of factors such as whether we are dealing with a crystalline, liquid or gas phase sample, whether the molecule is small or large, how symmetric it is and whether we need precise determination of bond lengths and bond angles. In cases where we need more general structural information, we might be interested, for example, in: the structural relationship between functional groups in an organic molecule; or the arrangements of ligands around a metal centre in a transition metal complex; or indeed in the way in which a cardiovascular drug molecule might bind within the protein cavity in haemoglobin.

Most of the methods described in this book rely on the interaction of photons or electrons with the molecule of interest. Those that use photons exploit whichever regions of the electromagnetic spectrum are appropriate to: the probing of rotational,

It is worth noting that mass spectrometry uses a wide variety of alternative ionization methods other than electron impact. These include fast atom bombardment, chemical ionization, electrospray ionization, matrix-assisted laser desorption/ionization, and inductively coupled plasmas. Similarly, diffraction experiments are not limited to X-rays and electrons but can also be conducted with neutrons.

vibrational, or electronic degrees of freedom in, respectively, rotational, vibrational, and electronic spectroscopy; the interaction of nuclear magnetic moments with externally applied magnetic fields in nuclear magnetic resonance spectroscopy (NMR); or the diffraction of X-rays in X-ray diffraction experiments. The two methods that largely employ electrons use them either as a means to transfer energy to ionize and fragment a molecule (mass spectrometry) or to exploit the wave-particle duality of their nature by observing their diffraction in either the gas or condensed phases.

1.2 Energy levels, transitions between them, and the electromagnetic spectrum

Visible light constitutes the part of the electromagnetic spectrum with which we are most familiar, simply because we are equipped with the facility to perceive it most acutely through our vision, but it forms just a small slice of the overall spectrum of light. Visible light assumes a particular historical importance in the development of spectroscopy because in sunlight we have a readily available source of radiation with which to observe certain phenomena. The dark lines discovered in the solar spectrum by Wollaston in 1802 and subsequently rediscovered by Fraunhofer 15 years later were due to absorption of sunlight by the colder gases in the outer regions of the Sun. The realization that discrete line spectra arise from discrete quantum states and, moreover, that the positions of those lines could provide a means to identify which atoms might be present, provided the basis of spectroscopic identification and characterization of gases, not only in remote bodies such as the Sun but also closer to home in our own atmosphere. Similarly, in emission rather than absorption, we can use colour as a means to identify particular elements in vaporized samples in flame tests. The characteristic yellow colour of a sodium flame, visible for example when you place a sample of common table salt into a flame, precisely corresponds in wavelength to one of the dark Fraunhofer lines, thereby identifying sodium as being present in the Sun.

The development of quantum theory, and subsequently of quantum mechanics, formed the basis of our understanding of the relationship between differences in energy between quantum states and the positions of lines appearing in an absorption or emission spectrum. Electromagnetic radiation consists of oscillating electric and magnetic fields that can propagate through space. In a vacuum, all electromagnetic radiation travels with the same speed, 2.997×10^8 m s^{-1}. The oscillations associated with different types of electromagnetic radiation can be described simultaneously in terms of their *wavelength* (the distance between successive peaks or troughs in the wave) or their *frequency* (the number of complete wavelengths passing a given point per second). The relationship between the wavelength, λ, and the frequency, ν, is given by

$$c = \lambda \times \nu \tag{1.1}$$

where c is the speed of light in m s^{-1}, and the wavelength, λ, and frequency, ν, are expressed, respectively, in units of length, m, and reciprocal time, s^{-1} or Hz (Hertz). Figure 1.1 shows how wavelength and frequency are inter-related.

The connection between the frequency of light and its ability to change the *energetic* state of matter was provided by Planck at the start of the 20th century when he proposed that frequency is related to energy, E, through the relationship

$$E = h\nu \tag{1.2}$$

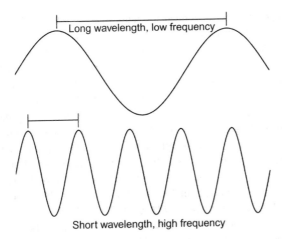

Figure 1.1 Electromagnetic radiation can be characterized in terms of both its wavelength and frequency. The wavelength provides a measure of the distance between successive peaks or troughs in the wave whilst the frequency provides a measure of the number of complete wavelengths passing a given point per unit time. The energy of the radiation scales linearly with the frequency.

where h is Planck's constant 6.626×10^{-34} J s. Effectively, the energy is delivered in discrete packets, $h\nu$, which we call either quanta or, to use the term coined by G.N. Lewis, photons. Thus, spectroscopy concerns the absorption, emission or scattering of photons by atoms and molecules, with those processes providing direct information about changes in internal quantum state. In the context of absorption or emission, the energy of the photon, E, supplied or generated by the process has to be exactly the same as the *difference* in energy, ΔE, between the two states involved (see Figure 1.2).

The energy states of an isolated atom are defined solely by the relationship between the orbiting electrons and the positively-charged nucleus. The translational freedom of the atom in space is not one which results in discrete quantum states because the atom's movements are not restricted in any way. Thus, the quantum states of an atom result solely from the fact that the motion of the electrons is constrained by the influence of the nucleus and, consequently, excited states can only result from the promotion of an electron from one orbit to another. To a first approximation then, an absorption spectrum in an atom has a somewhat simple structure involving a series of lines whose separation decreases rapidly as the electron becomes more and more weakly associated with the nucleus, eventually converging to a continuum at the point of ionization.

In a diatomic molecule, in addition to the constraints imposed upon the electrons, the translational motions of the two nuclei are now also constrained, in this case by the 'electronic glue' provided by the chemical bond between the nuclei. Certainly, the molecule as a whole can translate through space but this motion, like that of the isolated atom, is not one which results in stationary quantum states. However, movement of the two nuclei with respect to each other leads to two additional types of motion: rotation about an axis perpendicular to the internuclear axis, and motion of the two nuclei against one another along the internuclear axis. The former is known as a rotational degree of freedom, whilst the latter motion is known as a vibrational degree of freedom.

In addition, when certain nuclei are subjected to an externally applied magnetic field, their energy levels split to an extent which depends on the strength of the field, but this

The almost vanishingly small magnitude of Planck's constant goes some way to explaining why the quantization of energy in microscopic systems escaped notice for so long!

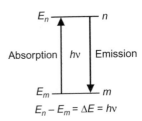

$$E_n - E_m = \Delta E = h\nu$$

Figure 1.2 Transitions between states m and n requires light to be either absorbed or emitted whose energy exactly equals the energy difference between the two states.

In practice, atomic spectra can be hugely complex. This complexity is most apparent in multi-electron atoms and is the result of electron–electron interactions involving orbital and spin angular momenta as well as interactions between the angular momenta associated with the electrons and those associated with the nuclei.

The wavenumber unit, cm^{-1}, $\tilde{\nu}$, is simply the reciprocal of the wavelength expressed in cm, and is commonly used by spectroscopists as a matter of convenience when working in certain parts of the electromagnetic spectrum. More generally, you will find that some scientific disciplines favour wavelength, and some frequency or wavenumber, both of which scale linearly with energy. In each case, the preferred units will depend largely on the magnitude of number resulting from their use but also in some cases simply as a consequence of historic convention.

splitting is of a very small order indeed and requires energy in the radiofrequency part of the electromagnetic spectrum to probe. This is the basis of NMR spectroscopy.

These principles extend to molecules of all shapes and sizes. All of them exhibit discrete line spectra as a result of the constraints imposed upon the motions of their constituent nuclei, on the electrons that form the bonds between them and as a result of perturbations provided by externally applied fields. We can get a sense of the relationship between the energy level spacings between electronic, vibrational, rotational, and nuclear magnetic energy levels by referencing them to the electromagnetic spectrum, presented in terms of wavelength, wavenumber and frequency (see Figure 1.3). So let's take a wander through the spectrum to orient ourselves and to get a sense of which types of light excite particular degrees of freedom in the molecule.

Visible light extends from about 390 to 770 nm, which in frequency terms spans about 7.7×10^{14} to 3.9×10^{14} Hz (i.e. 770 to 390 THz). With frequencies so large, you can see why spectroscopists prefer working in units of wavenumber, cm^{-1}, for which the same region covers about 26000 to 13000 cm^{-1}. Visible light is generally used to excite transitions between electronic energy levels, typically those in which the electrons are moving between valence orbitals.

Moving to longer wavelengths, lower energy, and hence lower frequencies, we find ourselves in the *infrared* region which covers 770 nm to about 1000 μm. In frequency terms this spans 390 THz to 300 GHz or, in wavenumbers, 13000 to 10 cm^{-1}. Infrared

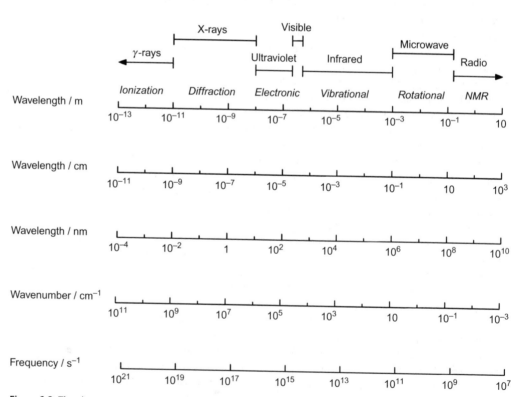

Figure 1.3 The electromagnetic spectrum extends from γ-rays and X-rays at the very short wavelength, high energy end of the spectrum to microwaves and radiowaves at the very long wavelength, low energy end. We can see from the figure that the visible region actually constitutes only a very narrow slice of the overall spectrum.

light is used to excite transitions between vibrational energy levels and, for the smallest and lightest molecules, rotational energy levels too. To longer wavelength still, we arrive first in the *millimetre* wave region (1 mm to 10 mm or 300 GHz to 30 GHz), and then *microwave* (10 mm to 30 cm or 30 GHz to 1 GHz) both of which are used in rotational spectroscopy. The long wavelength extreme of the electromagnetic spectrum brings us finally into the radiowave region (30 cm to about 10 m or 1 GHz to 30 MHz), which is used most notably in radio astronomy to look at the lowest energy rotational transitions, and, of course, in NMR spectroscopy.

We know from frequent warnings about the dangers of excessive sun bathing, or the fact that our dentist hides behind a lead shield when we are having our teeth X-rayed, that shorter wavelength light is more energetic than visible and, therefore, potentially damaging to human tissue. Beyond visible light to shorter wavelengths we travel first through the near *ultraviolet* (about 200 nm or 50000 cm^{-1}), into the far or vacuum ultraviolet (extending to 10 nm or 1×10^6 cm^{-1}) and then on to *X-rays* and *γ-rays* in the most energetic part of the spectrum. Near-UV is used in electronic spectroscopy, vacuum UV in experiments designed to probe very highly excited electronic states as well as in photoelectron spectroscopy, and X-rays in core photoelectron spectroscopy but more pertinent to this book, in X-ray diffraction.

The wide range of tools at our disposal illustrates that we can adopt many diverse approaches to obtain information about molecular structure. We can get a real sense of how the masses in a molecule are distributed by observing how it behaves as it rotates in space. We can infer a molecule's construction, size, and the presence of functional groups by seeing how the nuclei vibrate against one another. We can weaken or strengthen the bonds by exciting electrons to provide clues about the types of chromophores present in a molecule. A molecule may be broken into pieces and we can measure the masses to identify key constituent parts. We can use a molecule to diffract X-rays or electrons, thereby gaining a more holistic view of its structure. Finally, we can use the interaction of magnetic fields with nuclear spin moments to obtain information about the chemical environment of specific nuclei. None of the techniques described in this book provides a universal solution to structure determination in isolation but when used together in complementary combinations, become essential tools in the chemist's arsenal.

A chromophore is a group in a molecule with a characteristic optical absorption commonly responsible for its colour.

We have chosen to organize this book by looking initially, in Chapter 2, at the use of high-resolution spectroscopic techniques: microwave, infrared, and Raman spectroscopies can provide very precise determinations of molecular parameters such as bond lengths of molecules, typically at the smaller end of the scale. We then move on in Chapters 2 and 3 to consider how infrared and electronic spectroscopy can be used to obtain information about the structure of larger molecules through the identification of functional groups or chromophores and to provide insights into the symmetric properties of molecules. In Chapter 4, we discuss NMR and its ability to provide information about the local chemical environment of specific magnetic nuclei whilst in Chapter 5 we look at how mass spectrometry provides an unambiguous view of the overall mass of a molecule but critically also of its key structural components through fragmentation patterns. Both NMR and mass spectrometry can give insights into molecules ranging in size from small to very large indeed. In the final chapter, we look at the use of diffraction methods to elucidate complete structures of, typically, molecules at the larger end of the scale. Throughout the book we present examples of the applications of each technique in fields as diverse as radio astronomy, chemical analysis, materials science, biology, and medicine.

Rotational and vibrational spectroscopy

2.1 Introduction

All molecules are formed from collections of atoms bound together in some sort of geometrical framework. For molecules that are less constrained by the strong inter-molecular interactions that exist in the solid state, it is not unreasonable to speculate that we might learn something about the arrangement of those atoms by studying how the molecule behaves as it rotates in space and how it vibrates. We can make easy analogies to everyday experience by thinking, for example, about the ice-skater who spins more rapidly as he draws his arms close to his body; or the different sonic signature of a violin compared to a piano. We can exploit both concepts on the molecular scale by exploring how light interacts with the energy levels associated with rotational and vibrational degrees of freedom.

This chapter will present a brief introduction to rotational and vibrational spectroscopy, focusing initially on pure rotational spectroscopy and how it can be used not only in the precise determination of molecular geometry in small molecules but also in providing a direct means to deduce the shape of larger molecules. In the second part of the chapter, we will look at how vibrational spectroscopy can be used in the measurement of geometrical parameters of gas phase molecules, in providing information about the structure and shape of molecules in the condensed phase as well as in discussing the important role it plays in characterization and analysis in organic chemistry where it is used typically in tandem with mass spectrometry and NMR spectroscopy.

2.2 Rotational spectroscopy

Anyone who has thrown an unevenly proportioned stick for a dog to fetch might appreciate how its wobbly rotational arc through the air is defined by the way in which its mass is distributed along its length. You get some real sense of the extent to which its centre-of-mass lies away from its geometrical centre both from the feeling you get in throwing the stick (this will depend on whether you had grasped the heavier end or the lighter end) and from the way it subsequently cuts through the air. Similarly, a fidgety tennis player, nervously spinning her racket will notice that it spins rapidly and easily along the axis containing the handle but if attempting a nonchalant flip of the racket, releasing the handle and waiting for it to return to her grip after a single rotation, she will notice that its rotational speed is much more sedate and that it takes more effort to set that rotation in motion.

Both examples illustrate the principle behind rotational spectroscopy: the way a molecule behaves as it rotates in space can be used to deduce something about its shape and structure. Quantum mechanically, that behaviour is revealed through the rotational energy level structure, which we interrogate by somehow persuading the molecule to jump between its rotational energy levels. As we shall see in the sections that follow, if that molecule is small and symmetric, then we can extract really quite precise measurements of key molecular parameters. For larger molecules, the rotational spectrum may reveal less finely grained information about the molecular structure but nevertheless can provide important information about the molecule's shape.

In studying the pure rotational energy level structure of a molecule, we can exploit one of two spectroscopic principles, one employing conventional optical absorption of light and the second employing the inelastic scattering of light.

Rotational energy levels are very closely spaced and in order to excite transitions between them using optical absorption, we need to employ radiation spanning the *microwave* to *far infrared regions* of the electromagnetic spectrum (30 cm to 1000 μm which in frequency terms covers the range 1 to 300 GHz). This is the basis of the technique known as *microwave spectroscopy* (or millimetre wave spectroscopy when working in the region between about 1 to 10 mm or 300 to 30 GHz) which can trace its history back to the period immediately following the Second World War when pioneering spectroscopists exploited the easy availability of the klystron microwave source which had been developed and used for RADAR applications.

The second approach somewhat counterintuitively excites transitions between rotational energy levels using visible photons, whose energy far exceeds that required to span the very small gaps between the rotational states. The technique is known as *rotational Raman spectroscopy* and, rather than relying on optical absorption, uses instead the principle of inelastic scattering, in which a collision between the photon of light and the rotating molecule leaves the photon with a little less energy and the molecule with a little more (or *vice versa*). Both microwave spectroscopy and rotational Raman spectroscopy can produce beautiful, exquisitely detailed spectra from which we can extract structural parameters with high precision. The magnitude of the spacings between the lines that we see in either a microwave spectrum (top right in Figure 2.1) or a rotational Raman spectrum (Figure 2.1 lower) allows us to obtain rotational constants which yield geometric parameters such as bond lengths and bond angles (see sections 2.3 and 2.4).

The scattering of light occurs when radiation is caused to deviate from a linear path through an interaction with any type of nonuniformity in the medium through which it is travelling. Scattering may occur from a collision with a particle of some sort, such as an atom or molecule, and if that happens then the scattering can be either elastic, in which case no energy is exchanged as a result of the collision or inelastic in which energy *is* exchanged. In an inelastic collision between a photon and a molecule, the photon may either give up a little bit of its energy to the molecule or the molecule may give up a little bit of its energy to the photon.

2.3 Rotational energy levels

The presence of discrete lines in any absorption spectrum (such as that in the rotational spectrum shown top right in Figure 2.1) directs us to consider the molecule's energy level structure and the transitions between energy levels that give rise to the spectrum. In the case of rotational energy levels our starting point must be to consider the rotational behaviour of a diatomic molecule.

The simplest model of a rotating diatomic molecule is the free rotor in which the two nuclei are assumed to be point-like masses, m_1 and m_2, separated by a fixed distance, r (see Figure 2.2). Classically, the energy of rotation is given by

$$E = \tfrac{1}{2}I\omega^2 \tag{2.1}$$

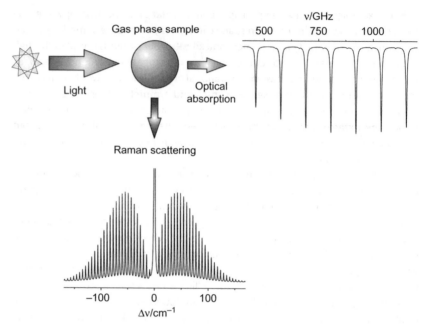

Figure 2.1 Changes in rotational state can be accomplished either through optical absorption of light in the microwave to far infrared regions of the electromagnetic spectrum (for polar molecules, top) or by inelastic scattering of, typically, visible light (bottom). This latter process is known as Raman scattering.

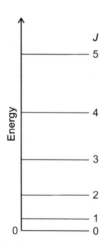

Figure 2.2 The simplest model of a diatomic molecule is that of the rigid rotor.

Figure 2.3 The energy levels of a rigid rotor increase quadratically with the rotational quantum number, J.

where ω is the angular velocity and I is the moment of inertia which is related to the masses of the nuclei and their separation through

$$I = \frac{m_1 m_2}{m_1 + m_2} r^2 \tag{2.2}$$

However, according to quantum theory, the energy (in Joules) of a molecular-scale rigid rotor cannot take any arbitrary value related to a seamlessly variable angular velocity, but rather takes discrete values given by the expression

$$E = \frac{h^2 J(J+1)}{8\pi^2 I} \tag{2.3}$$

where J is the rotational quantum number, which can take values 0, 1, 2, 3, ... and h is Planck's constant (6.626×10^{-34} J s). We can see from this expression that the energy takes discrete values that increase quadratically with J. This means, somewhat counter-intuitively, that the spacing between adjacent energy levels increases rapidly with every step up the energy level ladder (see Figure 2.3).

You may be thinking at this point that if the energy level spacings are related to the moment of inertia, I, and I is related to the distance separating the nuclei, then if we can measure the energy level spacings, we have a means to measure the distance, r. If you were thinking this, then your reasoning is spot on and moreover provides the basis on which rotational spectroscopy can be used in the precise determination of molecular structure.

2.4 Pure rotational microwave and millimetre wave spectroscopy of diatomic molecules

According to classical physics, light is generated as a consequence of a rapidly changing dipole moment. In the case of a rotating molecule, radiation of light associated with that motion can only occur if that molecule possesses a permanent dipole perpendicular to the axis of rotation. For a diatomic molecule, this requires the two nuclei to be different, in which case the molecule will have a permanent dipole in the direction of the internuclear axis. If we imagine the molecule rotating about an axis perpendicular to the internuclear axis, then it should be clear that the component of the dipole along a fixed direction will vary in proportion to the frequency of rotation (see Figure 2.4).

Classically, a rotating molecule will emit radiation at a frequency which equals its rotational frequency. It follows that the reverse must also be true, which is to say that molecules possessing a permanent dipole may absorb light which can then excite rotational motion if its frequency lies within the range of typical rotational frequencies for that molecule. These principles extend to the quantum world but in this case, rather than there being a seamless and continuous exchange of energy, the rotational energy of a heteronuclear molecule can only change if the molecule either absorbs or emits a photon of light having an energy which precisely matches the energy difference between two rotational energy levels. If we divide Equation (2.3) by hc on both sides, where h is Planck's constant and c is the speed of light in cm s^{-1}, we convert it from a *rotational energy level expression*, in which the units are J, into a *rotational term expression* with units of wavenumber, cm^{-1}.

$$F(J) = \frac{E}{hc} = \frac{h}{8\pi^2 Ic} J(J+1) = BJ(J+1) \text{ cm}^{-1} \qquad (2.4)$$

The constant,

$$B = \frac{h}{8\pi^2 Ic} \qquad (2.5)$$

is known as the rotational constant (sometimes referred to as the reciprocal moment of inertia).

Selection rules

The rotational energy state of a molecule is changed when radiation is absorbed or emitted that has an energy which equals the difference in energy between two rotational levels. The difference in wavenumber between the two levels is given by

$$\tilde{v} = F(J') - F(J'') = BJ'(J'+1) - BJ''(J''+1) \qquad (2.6)$$

where $F(J')$ is the term value of the upper rotational state and $F(J'')$ is the term value of the lower rotational state.

In practice, a molecule is not free to jump between any two rotational states, but instead this process is governed by the optical selection rules which require that J can only change by one unit at a time. We formally express this selection rule as

$$\Delta J = \pm 1 \qquad (2.7)$$

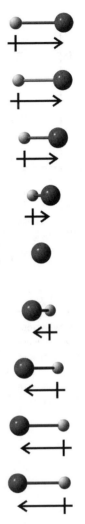

Figure 2.4 The magnitude of the component of the dipole of a rotating heteronuclear diatomic molecule in a fixed direction will vary in proportion to the frequency of rotation.

We have performed the transformation from *rotational energy level* expression to *rotational term* expression simply because the spectroscopic unit of the wavenumber is much more convenient to work with when dealing with the small quantities of energy involved in rotational and vibrational spectroscopy. For example, the rotational constant for the molecule H^{35}Cl is about 2.065×10^{-22} J but in wavenumbers it is a much friendlier 10.4 cm^{-1}.

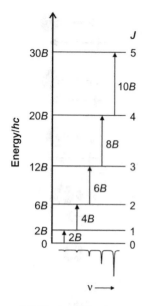

Figure 2.5 The transitions in absorption between rotational energy levels for a rigid rotor are governed by the $\Delta J=+1$ selection rule leading to a spectrum characterized by a series of lines equally spaced by twice the rotational constant, $2B$.

For the absorption of a photon we can further refine this to $\Delta J = J' - J'' = +1$ and so we can rewrite Equation (2.6) as

$$\tilde{v} = F(J''+1) - F(J'') = B(J''+1)(J''+2) - BJ''(J''+1)$$
$$= 2B(J''+1)$$

(2.8)

where J'' can take any integer value 0, 1, 2, 3, ... It is common practice to drop the ", replacing J'' with J, resulting in the slightly simplified expression

$$\tilde{v} = 2B(J+1)$$

(2.9)

which predicts a spectrum whose lines are equally spaced by twice the rotational constant, $2B$ (see Figure 2.5).

The selection rules for a diatomic or linear polyatomic molecule can be summarized as follows:

- The molecule must have a permanent dipole moment ($\mu \neq 0$)
- $\Delta J = \pm 1$

The first of these conditions will only be satisfied for heteronuclear diatomic molecules or for non-symmetric linear polyatomic molecules. Thus, molecules such as H_2, N_2, O_2 and Cl_2 will not display a pure rotational spectrum but HF, NO or CO will (see rotational spectrum of CO in Figure 2.6 and its analysis in Worked examples 2.1 and 2.2). Similarly, CO_2, being a symmetric linear triatomic molecule, has no permanent dipole and so does not show a pure rotational spectrum, whereas HCN or HC≡CF are both polar and exhibit classic linear molecule rotational spectra. First-row heteronuclear diatomic molecules have a low mass and consequently a small moment of inertia (Equation (2.2)) which means that their rotational constant, B, will be relatively large (Equation (2.5)) and they will then exhibit rotational spectra extending into the far infrared region (15 to 1000 µm or 20 THz to 300 GHz). Most linear polyatomic molecules, with smaller B values deriving from larger moments of inertia associated with their greater size, will tend to absorb in the millimetre or microwave regions of the electromagnetic spectrum.

Worked example 2.1

Question. The average separation between adjacent lines in the rotational spectrum of ^{12}CO shown in Figure 2.6 is 3.833 cm^{-1}. Use Equations (2.9) and (2.5) to estimate the CO bond length, r_{CO}.

Answer. According to Equation (2.9), the spacing between adjacent lines in a pure rotational spectrum of a linear molecule is $2B$. Thus an average spacing between lines of 3.833 cm^{-1} yields an approximate value for the rotational constant of 1.917 cm^{-1}. Rearranging Equation (2.5) allows us to express the moment of inertia, I, in terms of the rotational constant B as

$$I = \frac{h}{8\pi^2 Bc}$$

(2.10)

but we also know that

$$I = \mu r^2_{CO}$$

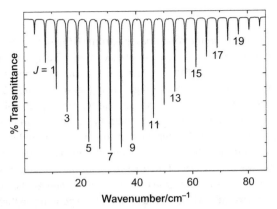

The transmittance, T, is the ratio of the intensity of transmitted light to the incident light intensity, given by $T = I/I_0$

Figure 2.6 The rotational spectrum of CO appears in the far infrared and exhibits the classic rotational structure of a linear molecule. The separation between adjacent lines is equal to twice the rotational constant, B. Note: J refers to J'', the rotational quantum number identifying the lower rotational level.

where

$$\mu = \frac{m_1 m_2}{m_1 + m_2} \qquad (2.11)$$

is known as the reduced mass. It follows that

$$r_{CO} = \sqrt{\frac{h}{8\pi^2 B \mu c}} \qquad (2.12)$$

The reduced molar mass of CO is

$$\mu = \frac{m_1 m_2}{m_1 + m_2} = \frac{12 \times 16}{12 + 16} = 6.857 \text{ g mol}^{-1}$$

which we convert to absolute reduced mass by dividing through by Avogadro's number to give

$$\mu = 1.139 \times 10^{-23} \text{ g}$$
$$= 1.139 \times 10^{-26} \text{ kg}$$

If we now substitute the values determined for the rotational constant and reduced mass into Equation (2.12), we obtain a value for the CO bond length of

$$r_{CO} = \sqrt{\frac{6.626 \times 10^{-34} \text{ J s}}{8\pi^2 \times 1.917 \text{ cm}^{-1} \times 1.139 \times 10^{-26} \text{ kg} \times 2.997 \times 10^{10} \text{ cm s}^{-1}}}$$
$$= 1.132 \times 10^{-10} \text{ m} = 0.1132 \text{ nm} = 1.132 \text{ Å}$$

Note: As we shall see later on in section 2.13, the bond length determined by this method is not the true equilibrium bond length, r_e, because the measurement is conducted with the molecule residing in its zero point vibrational level and not at the absolute minimum of potential energy. The difference between r_0, the average bond length in the zero point level, and r_e in most cases is very small but nevertheless, a determination of a bond length from a pure zero point level rotational spectrum will always only provide an approximation to r_e rather than a value of the very highest precision.

To a first approximation the spacings between adjacent rotational lines is equal. However, an increase in J means an increase in rotational energy, which corresponds classically to an increase in the frequency of rotation. You might give some thought to the consequences that this might have on the bond length and hence rotational constant (see Exercise 2.3).

If two or more states of a quantum mechanical system yield exactly the same measurable energy, then they are said to be degenerate. A good example of this that you might be familiar with is the idea that the three different 2p atomic orbitals in a first row atom all have exactly the same energy. The only difference between the three orbitals is their orientation in space, and in the absence of any means for the atom to tell which direction is which, all three of those orientations must be energetically equivalent.

In the absence of an external electric or magnetic field, a rotating diatomic molecule has no sense of direction and consequently, each rotational energy level is associated only with the amount of rotational energy in the system, regardless of how that rotation is oriented with respect to an external observer. However, if we apply an external field, the energy states of the system will now depend on how the fluctuating dipole, as perceived by an external observer (see Figure 2.4), is oriented with respect to the electric field direction. Quantum mechanically, there are $2J + 1$ possible ways in which a rotating molecule can orient itself relative to an arbitrarily defined axis and this then implies a $2J + 1$ degeneracy associated with each energy level in the absence of an external field. You do not need to worry about the details here because they lie outside the scope of this book but you do need to appreciate that in the quantum world, the quantization of physical properties is an inevitable consequence of any kind of constraint applied to the system.

In considering the population of a particular initial rotational level, we have also to take into account other processes which may be contributing to the steady-state population of that level. Although the most important contribution is that due to the thermal distribution, the populations of each level will also be influenced by optical absorption or emission processes which may deplete population in a particular level as well as emission from higher-lying levels which will contribute to an increase in population. To a first approximation, these additional factors may be ignored but should not be forgotten entirely!

Intensities

As you may have noticed in the rotational spectrum of CO shown in Figure 2.6, the intensities of the rotational lines increase markedly as we climb up the rotational energy level ladder and move from the low wavenumber part of the spectrum to higher wavenumbers, but then starts to decrease again to higher wavenumber still. The intensity of each spectral line depends, in part, on the transition probability, which can be calculated from the square of a mathematical quantity known as the transition moment integral. However, it also depends on the number of molecules in each of the various initial states and this is determined predominantly by the fact that the molecular sample will be in a state of thermal equilibrium. In rotational spectroscopy, because the quantum states are so close to one another in energy, the thermal energy is sufficient to populate quite a large number of states, depending on the temperature, which means that a rotational spectrum will contain quite a large number of lines.

According to the Maxwell–Boltzmann distribution law, the number of molecules in each quantum state is proportional to the Boltzmann factor $e^{-E/kT}$, where E is the energy of a particular level relative to the energy zero, k is the Boltzmann constant and T is the temperature in Kelvin. In the case of the vibrational energy levels of a diatomic molecule discussed later on in section 2.10, where there is generally no implied degeneracy associated with each level, the population N_v of the vth vibrational level relative to that of the zero point level, N_0, is given by

$$\frac{N_v}{N_0} = e^{-E_v/kT} \tag{2.13}$$

The implication here is that the greatest population will always be concentrated in the lowest energy quantum state, with the population in higher levels dropping off exponentially at a rate which depends on the magnitude of the spacing between the levels and the temperature. However, for rotational energy levels, we have also to account for a $2J + 1$ degeneracy which exists for each rotational energy level and we need to incorporate this into the expression to calculate the relative population of each level. Thus, the population N_J of the Jth rotational level relative to that of the lowest energy level, N_0, is given by

$$\frac{N_J}{N_0} = (2J+1)e^{-E_J/kT} \tag{2.14}$$

The $2J + 1$ factor in Equation (2.14) increases in size with J whilst the exponential term decreases: the result is that N_J/N_0 will increase up to the point that the exponential term wins the battle and takes over. This point occurs at a value of J given by

$$J_{max} = \left(\frac{kT}{2B}\right)^{1/2} - \frac{1}{2} \tag{2.15}$$

the value of J corresponding to the most populated level.

Worked example 2.2

Question. Use Equation (2.15) and the measured rotational constant for CO determined in Worked example 2.1 to calculate J_{max} at 298 K. Compare your result with the actual J_{max} obtained from the spectrum shown in Figure 2.6.

Answer. The rotational constant of ^{12}CO is 1.917 cm^{-1} and at 298 K, $kT = 4.114 \times 10^{-21}$ J. Converting kT from J to cm^{-1} yields a value for kT of 207.1 cm^{-1}. Inserting B and kT into Equation (2.15) yields a value for J_{max} of

$$J_{max} = \left(\frac{207.1}{2 \times 1.917}\right)^{1/2} - \frac{1}{2} = 6.85 \approx 7$$

The most intense line in the rotational spectrum shown in Figure 2.6 is the 8th line for which the lower rotational level is $J = 7$. This result is consistent with the spectrum having been recorded at room temperature.

2.5 Microwave and millimetre wave spectroscopy of linear triatomic molecules

From the perspective of an external observer, a rotating dipole associated with a linear polyatomic molecule looks no different from that of a rotating heteronuclear diatomic molecule. In both cases, the dipole is directed along the internuclear axis and consequently, both diatomic and linear polyatomic molecules only have a single measurable moment of inertia, the magnitude of which will depend upon the reduced mass of the system (sometimes referred to as the effective mass in molecules with more than two nuclei).

The rigid rotor model of a linear triatomic molecule (see Figure 2.7) is constructed by arranging three point masses, m_1, m_2, and m_3 in line, with their relative separations defined by two bond lengths r_{12} and r_{23}.

The obvious problem presented by such molecules is that their rotational spectra can only ever yield a single rotational constant associated with a single moment of inertia and yet we need two bond lengths in order to properly describe the molecular geometry. We can illustrate how we get around this problem by considering the story of the first rotational spectrum of a molecular ion.

In experiments conducted in 1970 using the 36 foot telescope of the National Radio Astronomy Observatory in Arizona, a strong millimetre wave emission was detected at 89190 MHz in a number of separate observations of interstellar clouds (see Figure 2.8).

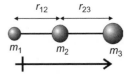

Figure 2.7 A rigid rotor model of a linear triatomic molecule features two structural parameters, r_{12} and r_{23}, which define the two bond lengths. However, the single moment of inertia obtained will yield only a single structural parameter.

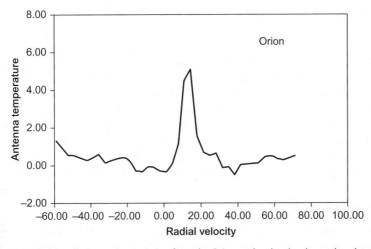

Figure 2.8 Strong millimetre wave emission from the Orion molecular cloud complex, detected at 89190 MHz using the 36 foot telescope of the National Radio Astronomy Observatory in Arizona. Reproduced from D. Buhl and L.E. Snyder, *Nature*, **228**, 267 (1970) with permission.

Up to this point we have only been considering optical *absorption* in which the energy supplied by the photon lifts the molecule into a higher rotational state. However, 'what goes up, must come down' and molecules in excited rotational states can emit radiation as they fall down to lower rotational levels. Both processes are governed by the same selection rules and can simply be thought of as the inverse of one another. The obvious advantage of microwave and millimetre wave *emission* measurements of remote species is that you do not need to supply any exciting radiation!

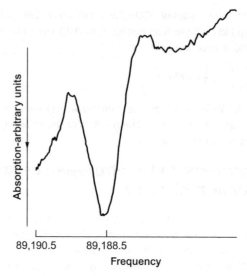

Figure 2.9 The $J = 0 \rightarrow 1$ transition in HCO$^+$ recorded in a liquid nitrogen cooled discharge of a CO-H$_2$ mixture at 10 mTorr. The spectrum was recorded in a microwave spectrometer using frequency modulation. Reproduced from R.C. Woods, T.A. Dixon, R.J. Saykally, and P.G. Szanto, *Physical Review Letters*, **35**, 1269 (1975) with permission. Copyright (1975) American Physical Society.

In spite of the accidental nature of the original astronomical observation, this discovery turned out to be very important because HCO$^+$ has since been recognized as the cornerstone species in the formation of interstellar molecules *via* ion-molecule reactions occurring in the interstellar medium. The abundance of molecular ions in interstellar clouds derives from the high flux of stellar radiation, cosmic rays and fast electrons and once produced, they persist because of the low molecular density in the clouds.

At the time, this mysterious line was unknown in laboratory experiments and, in the absence of a positive identification, the scientists undertaking these experiments named this new species X-ogen. A number of molecules with transitions in the neighbourhood of 89190 MHz (89.19 GHz) were suggested as possible candidates, including H$_2{}^{32}$S and HNC. In spite of the speculation about the possible identity of X-ogen, it was not until 5 years later that laboratory experiments confirmed the identity of the carrier as protonated CO, HCO$^+$. In these laboratory experiments, HCO$^+$ was made from a discharge of various mixtures of hydrogen and carbon monoxide cooled to near liquid nitrogen temperatures. The line observed in these experiments appeared at 89188.545 $\pm$ 0.020 MHz and was assigned to the $J = 0 \rightarrow 1$ transition in HCO$^+$ (see Figure 2.9).

The millimetre wave spectra of HCO$^+$ can only yield a single moment of inertia and consequently only a single structural parameter. How then do we obtain the two bond lengths, r_{CH} and r_{CO} from our experimental spectrum? Well, we can circumvent the problem by making use of the fact that molecular bond strengths and bond dimensions are defined by the electronic structure of the molecule and not by the masses of the nuclei. Consequently, an isotopically substituted HCO$^+$, such as DCO$^+$ or H^{13}CO$^+$ will have exactly the same equilibrium bond lengths as H^{12}CO$^+$ but because the effective mass and hence moment of inertia is now different, we can now use two measured moments of inertia from two isotopomers to obtain the two required structural parameters.

Worked example 2.3

Question. The $J = 0 \rightarrow 1$ transition in HCO$^+$ is observed at 89188.5 MHz whilst the corresponding transition in DCO$^+$ occurs at 72039.3 MHz. Calculate the rotational constant and moment of inertia in each case and hence determine the HC and CO bond lengths in HCO$^+$.

Answer. The moment of inertia for a linear triatomic molecule can be expressed conveniently in the following form:

$$I = \frac{1}{M}\left(m_1 m_2 r_{12}^2 + m_1 m_3 r_{13}^2 + m_2 m_3 r_{23}^2\right) \tag{2.16}$$

where M is the total mass of the molecule.

The $J = 0 \rightarrow 1$ transition frequencies for HCO$^+$ and DCO$^+$ yield rotational constants of 44594.25 and 36019.64 MHz, respectively. Inserting each of these values into

$$I = \frac{h}{8\pi^2 B},$$

recognizing that we are working in units of Hz (s^{-1}) now rather than cm^{-1}, we obtain two moments of inertia

$$I_{HCO^+} = 1.882 \times 10^{-46} \text{ kg m}^2$$

$$I_{DCO^+} = 3.330 \times 10^{-46} \text{ kg m}^2$$

It is rather more convenient at this point to convert these moments of inertia from S.I. units into atomic units yielding

$$I_{HCO^+} = 11.3328 \text{ u.Å}^2$$

$$I_{DCO^+} = 14.0306 \text{ u.Å}^2$$

where u is the unified atomic mass unit and Å is the atomic unit of length.

Substituting each of these moments of inertia in turn into Equation (2.16) together with the known masses of the individual nuclei, m_H, m_D, m_C, and m_O and the total masses, M_{HCO^+} and M_{DCO^+}, of the two isotopologues, HCO$^+$ and DCO$^+$, yields a pair of simultaneous equations

$$I_{HCO^+} = 11.3328 = \frac{1}{29.008}\left(12.094 \times r_{12}^2 + 16.125 r_{13}^2 + 192 r_{23}^2\right) \text{u.Å}^2$$

$$I_{DCO^+} = 14.0306 = \frac{1}{30.014}\left(24.168 \times r_{12}^2 + 32.224 r_{13}^2 + 192 r_{23}^2\right) \text{u.Å}^2$$

At face value, this might appear to present an insoluble problem because we are seeking to obtain three parameters, r_{12}, r_{13}, and r_{23}, from two equations but of course we recognize that r_{13} is simply the sum of r_{12} and r_{23} and so both equations can be expressed in terms of the same two parameters, r_{12} and r_{23}.

$$I_{HCO^+} = 11.3328 = \frac{1}{29.008}\left(12.094 \times r_{12}^2 + 16.125(r_{12} + r_{23})^2 + 192 r_{23}^2\right) \text{u.Å}^2 \tag{2.17}$$

$$I_{DCO^+} = 14.0306 = \frac{1}{30.014}\left(24.168 \times r_{12}^2 + 32.224(r_{12} + r_{23})^2 + 192 r_{23}^2\right) \text{u.Å}^2 \tag{2.18}$$

With a little bit of mathematical juggling, we can solve for both r_{12} and r_{23} (see Exercise 2.6).

The unified atomic mass unit is the standard unit, u, used to indicate mass on an atomic or molecular scale and is defined as one twelfth of the mass of an unbound carbon atom. Its value is $1.660538782(83) \times 10^{-27}$ kg which is approximately equal to the mass of a proton or neutron. The atomic unit of length is the Ångström, Å, with a value of 10^{-10} m (one ten-billionth of a metre). We must also remember that rotational spectroscopy is very sensitive to different isotopes and so in calculating bond lengths from moments of inertia, we need to remember to use the atomic masses for the specific isotopes we are interested in and not necessarily the values displayed in most periodic tables which present atomic masses weighted over the abundances of the naturally occurring isotopes. Thus, $m_H = 1.0078$ u, $m_D = 2.0141$ u, $m_C = 12.00$ u and $m_O = 16.00$ u

2.6 Rotational spectroscopy of non-linear polyatomic molecules

The preceding discussion of rotations of diatomic and linear polyatomic molecules has been presented in terms of a single moment of inertia related to a single rotational constant from which a single geometrical parameter derives. However, all molecules, regardless of their shape or complexity, actually have three moments of inertia, associated with three rotational degrees of freedom about three mutually perpendicular Cartesian axes. In the case of any linear molecule, one of these moments of inertia will be of zero value (that describing rotation about the internuclear axis) and the remaining two will be equal to one another ($I_b = I_c, I_a = 0$) and so consequently it is that single effective moment of inertia that determines the rotational energy level structure of a linear molecule.

$$x\ (c)$$
$$\uparrow$$
$$\text{H——C}\equiv\text{N}\longrightarrow z\ (a)$$
$$\downarrow$$
$$y\ (b)$$

The key to molecular structure determination from rotational spectroscopy lies in the relationship between the three moments of inertia, and consequently it becomes possible to classify molecules broadly speaking into different types of shape. A symmetric top molecule has two of its principal moments of inertia equal, with the third non-zero and may be either prolate, such as CH_3Cl for which $I_a \leq I_b = I_c$, or oblate, such as benzene in which case $I_a = I_b \leq I_c$. An asymmetric top has all three moments of inertia unequal $I_a \neq I_b \neq I_c$, a classic example of which is water.

However, it turns out that many examples of asymmetric rotors have two almost equal moments of inertia and these are then treated as either prolate near-symmetric rotors, for which $I_a \leq I_b \cong I_c$, or oblate near-symmetric rotors, in which case $I_a \cong I_b \leq I_c$.

A spherical rotor is one which has all three moments of inertia equal ($I_a = I_b = I_c$), such as SF_6 or methane. It might seem strange to be considering spherically symmetric molecules in a discussion of rotational spectroscopy because by definition a spherical rotor will have no permanent dipole and consequently will not display a rotational spectrum. This is certainly true of SF_6 but in tetrahedral molecules such as methane which are also spherically symmetric, rotation about the axis passing through one of the CH bonds will cause the other three CH bonds to extend under the influence of centrifugal forces and the molecule effectively becomes a symmetric rotor and will display a weak rotational spectrum.

A detailed discussion of the rotational energy level structure for each type of rotor is beyond the scope of this book but we can briefly consider how we might extract detailed information about the molecular structure of such molecules from the rotational spectrum. For a symmetric top, two of the three moments of inertia are the same and consequently a rotational spectrum will only yield two rotational constants, one associated with the unequal moment of inertia and the second with the pair of equal moments. Consequently, we can only ever determine two geometric parameters from the two rotational constants for a symmetric top. In the case of an asymmetric rotor, there will be three rotational constants but in a planar asymmetric rotor such as formaldehyde, the out-of-plane moment is equal to the sum of the other two and consequently, three rotational constants will again only ever yield two geometrical parameters.

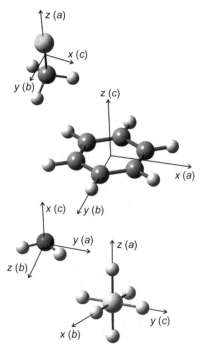

It is clear then, that while a rotational spectrum can provide very finely resolved spectral information, in general, rotational spectroscopy appears only to be able to provide rather coarse insight into the structural properties of molecules with more than two nuclei. However, as we saw in the case of HCO$^+$ in Worked example 2.3, we can use strategic isotopic substitution to obtain a greater number of geometrical parameters. With a great deal of effort, it is possible to obtain a complete description of the structure of quite large molecules using this approach but more generally the emphasis may be more about using the spectrum to deduce the type of structure and to then use selective isotopic substitution to reveal key geometric parameters. A good example of the latter approach is illustrated by a recent study of the halogen-bound complex $H_2O \cdots ICF_3$.

In that study, microwave spectra of four isotopologues of $H_2O \cdots ICF_3$ were recorded between 7 and 18 GHz. By comparing their experimental spectra with simulations generated using a symmetric top model, the authors were able to determine that the water molecule binds to the ICF_3 via an $O \cdots I$ halogen bond and that the water sits symmetrically astride the principal rotational axis of the ICF_3 with the two hydrogen atoms hanging down either side. They were then able to obtain values for the halogen bond length, ($r(O \cdots I) = 3.0517 \pm 0.0018$ Å) and of the angle ($\phi = 34.3 \pm 0.2°$) between the two-fold rotational axis of the water and the three-fold rotational axis of the ICF_3 (see Figure 2.10).

A halogen bond derives from a weak interaction involving a halogen atom as an acceptor of electron density. A halogen bond can thus be described as $D \cdots X\text{-}Y$, where D is an electron donor (Lewis base), X is the halogen (Lewis acid), and Y is typically carbon, nitrogen or another halogen. Direct analogies can be made with the more familiar hydrogen bond ($D \cdots H\text{-}Y$) in which it is the hydrogen atom that plays the role of electron acceptor. The halogen bond assumes a particular significance in crystal engineering where it is used in the design of crystalline materials with applications as diverse as liquid crystal development and hydrogen storage. Recent interest in self-assembling crystal structures which coordinate through halogen bonds has prompted interest in the study of model halogen-bonded systems composed of a simple Lewis base such as H_2O interacting with the electron-rich halogen atom in a small, halogenated iodoalkane.

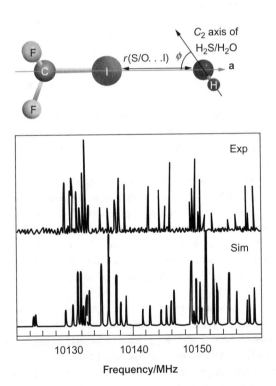

Figure 2.10 A section of the broadband microwave spectrum of $H_2O\cdots ICF_3$. The inset displays an expanded section. The simulated spectrum (bottom) uses the geometrical parameters determined for $H_2{}^{16}O\cdots ICF_3$. Reproduced from S.L. Stephens, N.R. Walker, and A.C. Legon, *Phys. Chem. Chem. Phys.*, **13**, 21093 (2011) with permission.

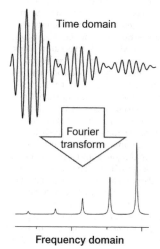

Time domain

Fourier transform

Frequency domain

Figure 2.11 A Fourier transform microwave spectrometer works by exposing the sample to a pulse of microwave radiation within a tuned cavity. This pulse causes all of the molecules in the spectrometer to rotate in phase with one another. Once the pulse is switched off, the microwave emission which accompanies the decay of this collective polarization (known as free induction decay) is recorded and a frequency domain spectrum obtained by performing a mathematical operation on the time domain signal known as a Fourier transform (this principle is very similar to that used in NMR spectroscopy described in Chapter 4). The microwave pulse is typically a fairly narrow band which means that each pulse will sample rotational transitions within a fairly narrow frequency window. A spectrum is acquired by tuning the cavity step by step across the frequency range of interest.

2.7 Experimental methods in microwave and millimetre wave spectroscopy

A microwave or millimetre wave spectrometer, as used to record spectra such as those discussed previously, requires a source of tuneable monochromatic radiation, a waveguide acting as a sample cell through which the radiation is passed, and a detector. Microwave radiation covers the range 1–30 GHz (300–1 mm) with millimetre wave radiation covering the range 30–300 GHz (10–1 mm). A spectrum is obtained by monitoring the intensity of transmitted radiation as a function of radiation frequency. In the pioneering days of microwave spectroscopy in the aftermath of the Second World War, the most commonly used source of microwave radiation was the *klystron* which had been developed and used in RADAR applications in the war. The klystron is essentially an evacuated cavity through which a beam of electrons is accelerated and reflected by a reflector held at high voltage. Electromagnetic oscillations build up in the cavity and modulate the electron beam so that the electrons bunch together. This bunching of electrons regenerates and amplifies the microwave radiation at a wavelength tuned to the dimensions of the cavity. The frequency of the klystron can be changed to a small degree by electrical tuning over a range of a few tens of MHz with coarser tuning achieved by deforming the cavity mechanically. Millimetre waves are less commonly used in rotational spectroscopy than microwaves but can be generated from the same sources by using frequency multipliers which generate harmonics of the longer wavelength microwave radiation. In the 1960s, a more convenient source of microwaves was developed known as the *backward wave oscillator*. This source is related to the klystron, but has the advantage that it can be tuned electronically over a wider frequency range rather than having to combine mechanical and electrical tuning of the klystron.

Although the general principle is simple enough, there are a number of technical difficulties to be overcome: the low energies of the light quanta make them much more difficult to detect than, say, infrared or visible light quanta; the molecular absorption cross-sections are very small in the microwave region which means that the loss in transmitted power resulting from absorption may be lower than 0.1% of the incident light intensity; the small separation between energy levels results in very small differences in Boltzmann population which means that using high microwave power can be counterproductive, leading to signal saturation. All of these issues as well as others, such as pressure and Doppler broadening, mean that microwave and millimetre wave spectroscopy are far more reliant on complex electronic devices than optical spectroscopy to separate the signals associated with absorption from noise.

Subsequent developments of microwave spectroscopy employed pulsed microwave sources combined with *Fourier transform techniques* (FTMW) to improve resolution and sensitivity; supersonic free jet expansion techniques to cool the sample down to a few Kelvin to concentrate the rotational population into the lowest rotational levels (and as a consequence permit the use of microwave spectroscopy to study weak interactions in gas phase clusters—illustrated in the example given in section 2.6); and with the huge advances in electronics in the past 10 years, the use of chirped microwave pulses to provide broadband capability to complement the high sensitivity and resolution of FTMW.

2.8 Rotational Raman spectroscopy

In spite of the fact that homonuclear diatomic molecules and symmetric linear poly-atomic molecules do not display pure rotational spectra, all hope is not lost. All molecules, regardless of whether they possess a permanent dipole, are polarizable when subject to an externally applied electric field: application of an electric field, F, will induce an electric dipole moment, $\boldsymbol{\mu}_{ind}$, as the positively charged nuclei are moved in one direction and the negatively charged electrons in the other (see Figure 2.13).

The magnitude of this induced dipole will depend not only on the strength of the electric field, the charge distribution, and density, but also, importantly, on how the molecule is oriented with respect to the field.

We can use this to our advantage by exploiting the fact that light contains an electric field component (the *electro* in electromagnetic radiation) which fluctuates in magnitude at the same frequency as the frequency of the light, ν. If we subject a molecule to such a light field, the induced oscillating dipole will itself radiate at the same frequency, ν, as the incident light. This radiation is known as *Rayleigh* scattering and is the result of photons being scattered *elastically*—having lost none of their original energy—following collision with the molecule.

As suggested above, the magnitude of the induced dipole depends on the orientation of the molecule with respect to the electric field which implies that the polarizability of the molecule—a measure of the extent to which the electrons can be polarized by an electric field—also depends on orientation. It follows therefore that the polarizability varies with the rotational motion of the molecule and an important consequence of this is that the molecule will radiate at three different frequencies; ν, the Rayleigh frequency; as well as $\nu - 2\nu_{rot}$ and $\nu + 2\nu_{rot}$. The two additional frequencies, $\nu - 2\nu_{rot}$ and $\nu + 2\nu_{rot}$, correspond to inelastically scattered light in which the light has either lost some energy, which is known as *Stokes* scattering, or gained some energy, which is known as *anti-Stokes* scattering (see Figure 2.14). This inelastic scattering is known as the Raman effect and in the context of rotational spectroscopy, the technique which exploits the effect is known as rotational Raman spectroscopy (see marginal note).

The result of the Raman scattering is that we expect to see displaced lines either side of the Rayleigh line at a distance of $2\nu_{rot}$. In quantum mechanical terms, for a linear molecule, this inelastic scattering leaves the molecule in a higher or lower rotational state, according to the selection rule

$$\Delta J = 0, \pm 2 \tag{2.19}$$

where the $\Delta J = 0$ line corresponds to the Rayleigh line, and the $\Delta J = \pm 2$ lines correspond to the Stokes and anti-Stokes lines, respectively (see Figure 2.15).

The outcome of this process may at first sight appear equivalent to optical absorption and emission but the important difference is that the Raman process can happen for any frequency of incident light, as long as its energy exceeds the energy gap needing to be bridged between states. Typically, Raman spectroscopy is conducted with visible photons whose energy vastly exceeds the energy gaps between adjacent rotational energy levels.

As with rotational spectroscopy discussed earlier, transitions may occur from any initially populated rotational state, but in this case governed by the $\Delta J = 0, \pm 2$ selection rule for linear molecules. Consequently, rather than a single long sequence of rotational lines seen in a pure optical rotational spectrum (such as that shown in Figure 2.6),

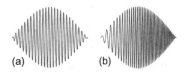

Figure 2.12 (a) A conventional FTMW experiment uses a microwave pulse having a fairly narrow bandwidth, i.e. the microwave frequency of the pulse does not change a great deal between the start and end of the pulse. (b) A chirped pulse is one in which the frequency of the pulse is ramped so that its frequency at the end of the pulse is much higher than its frequency at the beginning of the pulse. The huge advantage of such a pulse is that it allows a genuine broadband spectrum to be acquired within a single data acquisition shot, hugely decreasing data acquisition times.

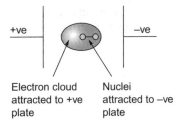

Electron cloud attracted to +ve plate   Nuclei attracted to −ve plate

Figure 2.13 The application of an electric field to a non-polar molecule will induce an electric dipole, the magnitude of which depends on the size of the electric field and the orientation of the molecule with respect to the field.

For visible or ultraviolet light, the nuclei are not able to respond fast enough to the oscillating field produced by the electric field component of the light and so the induced dipole results only from the oscillation of the electrons in the molecule and not of the nuclei.

The inelastic scattering of photons is analogous to Compton scattering of X-rays and was first discovered in the 1920s by a young Indian physicist named Chandrasekhara Venkata Raman, who was working at the University of Calcutta. The technique which he developed is known as Raman spectroscopy and is better known for its application in vibrational spectroscopy (see section 2.12) but the underlying process is exactly the same in both applications.

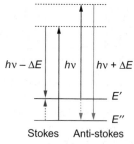

Figure 2.14 In the Raman effect, incident photons may scatter elastically, with no loss of energy (Rayleigh scattering) or inelastically where the photon may either lose energy (Stokes scattering) or gain energy (anti-Stokes scattering). The corresponding outcomes are: (i) the molecule remains in its initial quantum state (the most likely scenario); (ii) is raised to a higher quantum state; or (iii) is lowered to a lower quantum state. The dotted arrows indicate the actual transitions promoted by the process.

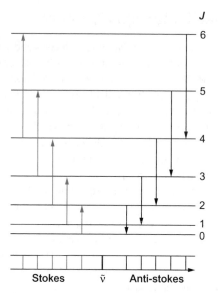

Figure 2.15 In a rotational Raman experiment, the transitions are governed by the $\Delta J = 0 \pm 2$ selection rule. The heavy line in the middle of the spectrum is the Rayleigh line whose wavenumber is undisplaced from that of the incident radiation. The lines to lower wavenumber are the Stokes lines and the lines to higher wavenumber are the anti-Stokes lines.

a rotational Raman spectrum will contain many lines distributed either side of a single intense Rayleigh line.

The magnitude of the wavenumber shift for both Stokes and anti-Stokes transitions is given by

$$\tilde{v} = F(J+2) - F(J) = B(J+2)(J+3) - BJ(J+1) \tag{2.20}$$
$$= 4BJ + 6BJ = 4B(J + \tfrac{3}{2})$$

where J refers to the lower state. We can see from this that the first Stokes line and the first anti-Stokes line, for which the lower level in each case is $J = 0$, will be displaced from the central Rayleigh line by $-6B$ and $+6B$, respectively. The subsequent lines appear at intervals of $4B$. This characteristic appearance can be seen quite clearly in the rotational Raman spectrum of $^{19}F_2$ shown in Figure 2.16.

One unexpected feature of the spectrum shown in Figure 2.16 is the alternating intensity of the rotational lines. A detailed explanation for this phenomenon is beyond the scope of this book but derives from the fact that the ^{19}F nucleus has a nuclear spin of ½ and in a homonuclear diatomic molecule such as $^{19}F_2$, the interaction of the nuclear spins of the two ^{19}F nuclei results in two nuclear spin isomers that lead to differences in the statistical weighting of populations of different J states (see also Chapter 4). Consequently, in this particular example, transitions originating in odd J levels have three times the intensity of those originating in even J levels.

Table 2.1

J	$|\tilde{v}| / cm^{-1}$
0	5.300
1	8.833
2	12.366
3	15.900
4	19.433
5	22.966
6	26.499
7	30.033
8	33.566
9	37.099

Worked example 2.4

Question. The positions of the first ten lines in the Stokes branch of the rotational Raman spectrum of $^{19}F_2$ shown in Figure 2.16 are given in Table 2.1. Use Equation (2.20) to find B and hence calculate an approximate bond length, r, for $^{19}F_2$.

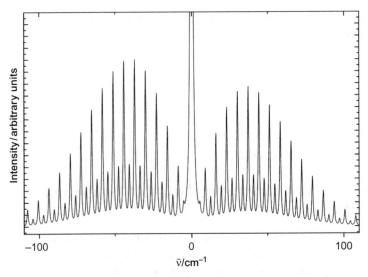

Figure 2.16 The rotational Raman spectrum of $^{19}F_2$. The intense feature at 0 cm^{-1} is the Rayleigh line, with the Stokes and anti-Stokes bands spreading to lower and higher relative wavenumber, respectively. The first member of each series is displaced by $6B$ from the Rayleigh line, with subsequent lines appearing at intervals of $4B$ thereafter.

Answer. According to Equation (2.20), the magnitude of the wavenumber shift for the Stokes transitions is given by $\tilde{\nu} = 4B(J+\frac{3}{2}) = 4BJ + 6B$. Thus, a plot of $\tilde{\nu}$ against J should yield a straight line plot with a gradient of $4B$ and a y-axis intercept at $6B$. Such a plot is shown in Figure 2.17.

The slope of the plot yields a value for $4B$ of 3.533 cm^{-1} and hence for the rotational constant B of 0.883 cm^{-1}. From here, we can proceed as in Worked example 2.1.

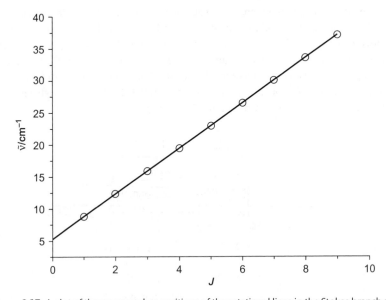

Figure 2.17 A plot of the wavenumber positions of the rotational lines in the Stokes branch of the rotational Raman spectrum of $^{19}F_2$.

With the rotational constant determined, the bond length of F_2 is straightforwardly calculated from

$$r = \sqrt{\frac{h}{8\pi^2 B\mu c}}$$

Substituting in the measured value for B and the reduced mass of $^{19}F_2$ yields

$$r = \sqrt{\frac{6.626 \times 10^{-34} \text{ J s}}{8\pi^2 \times 0.883 \text{ cm}^{-1} \times 1.578 \times 10^{-26} \text{ kg} \times 2.997 \times 10^{10} \text{ cm s}^{-1}}}$$
$$= 1.418 \times 10^{-10} \text{ m} = 1.418 \text{ Å}$$

2.9 Experimental methods in rotational Raman spectroscopy

The underlying experimental requirements for a Raman experiment are relatively straightforward but the method does present particular technical challenges that are related to the inherent weakness of the Stokes and anti-Stokes Raman scattering. The inefficiency of the process can be countered by the use of a very intense, highly mono-chromatic light source, with the scattered light being dispersed and detected normal to the direction of the incident radiation to avoid the incident radiation reaching the detector.

In early Raman experiments, the most intense sources of monochromatic light were atomic emission sources, with a discrete line in the visible or near ultraviolet being isolat-ed by optical filtering. The most commonly used such source was the mercury discharge lamp which produces three intense lines at 253.7 nm, 404.7 nm, and 435.8 nm. Such sources were combined with multi-pass absorption cells employing concave mirrors both to reflect the incident light back and forth and to maximize the collection efficiency of the scattered light (see Figure 2.18).

From the 1970s, atomic emission sources were replaced by lasers which pro-vided a more coherent and intense source of radiation but also much narrower line widths which allowed for better spectral resolution. Helium–neon (632.8 nm) and argon (514.5 nm) lasers were commonly used as sources of visible light in Raman experiments with Nd-YAG lasers providing near infrared light (1064 nm) for use with coloured samples. Although the scattering of visible light is far more efficient than infrared, molecules which absorb in the visible may fluoresce, with that fluorescence

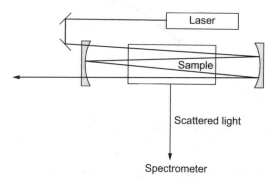

Figure 2.18 A multi-pass sample cell in a Raman spectrometer.

then swamping the Raman signal. The weakness of infrared scattering was overcome following developments in the application of Fourier transform techniques (described earlier in the context of microwave spectroscopy) to infrared spectroscopy as well as the use of much more sensitive semiconductor infrared detectors.

2.10 Vibrational spectroscopy

All molecules, regardless of their size or structural complexity, possess natural resonant frequencies of vibration in much the same way that each string of a bass guitar has its own natural frequency or in the way that Big Ben, the bell suspended from the clock tower of Westminster, has its own very distinctive resonant frequency. In fact, in our perception of the sounds emerging from any resonant musical instrument, or indeed the differences in the sound that a choral singer might make in York Minster compared to their bathroom, we get immediate insight into the size and shape and construction of the space in which the sound is resonating. Similarly, we can use the ways in which molecules vibrate to gain direct insight into their construction, their shapes and their geometries.

For simple diatomic molecules, with just two nuclei and one bond, there is just one such natural resonant frequency, commonly referred to as the normal mode of vibration, but for non-linear molecules with more than two nuclei, the number of modes of vibration increases according to 3N–6 where N is the number of nuclei. Thus, for example, water, with 3 nuclei, will have 3 normal modes of vibration, methane has 6 and benzene, with 12 nuclei, has 30! Normal modes of vibration can conveniently be thought of as being associated with nuclear displacement along a particular internal coordinate. In the case of water, the three normal modes correspond straightforwardly to a pair of bond stretching modes, one in which the two OH bonds simultaneously stretch and compress together (the symmetric stretch) and another in which one bond stretches whilst the second compresses (the antisymmetric stretch). The third mode is the bending mode for which the internal coordinate involves a change in bond angle rather than bond length (see Figure 2.19).

Linear molecules behave slightly differently in that they have one more vibrational mode than an equivalently sized non-linear molecule. Thus CO_2, with three nuclei arranged in a linear structure has 3N–5 = 4 vibrational modes (but only 3 vibrational frequencies because two of the modes have the same frequency i.e. they are degenerate).

When we think about vibrational frequencies, we might intuitively think in terms of the sorts of frequencies we encounter every day through our perception of sound. The range of frequencies audible to a young person extends from about 20 Hz at the low frequency end (the lowest pitched notes emanating from the pipes of a church organ) to about 20 kHz at the treble end (the high-pitch whine of a mosquito). Molecules, however, vibrate at frequencies which boggle the mind. For example, the nuclei in the N_2 molecule vibrate at a frequency of 7.1×10^{13} Hz—that is, at about seventy trillion

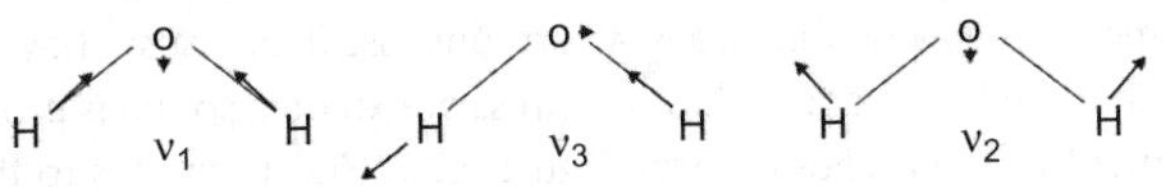

The number of vibrational modes for a molecule is obtained by subtracting the number of possible translational degrees of freedom (3 for both linear and non-linear molecules) and the number of rotational degrees of freedom (3 for a non-linear molecule but only 2 for a linear molecule) from the total number of degrees of freedom, 3N. A linear molecule only has two rotational degrees of freedom because rotation about the internuclear axis does not involve a moment of inertia—there is no mass being swung about.

Figure 2.19 The three normal modes of water, v_1, v_3 and v_2, are named the symmetric stretch, the antisymmetric stretch and the bending mode.

20 kHz=2×10^4 Hz=twenty thousand Hz

20 MHz=2×10^7 Hz=twenty million Hz

20 GHz=2×10^{10} Hz=twenty billion Hz

20 THz=2×10^{13} Hz=twenty trillion Hz

vibrations per second. If we convert this value from units of Hz (s^{-1}) into units of wavelength, by dividing into the speed of light, c, then we obtain a value of 3896 nm which is a wavelength commonly associated with that of light in the short wavelength infrared region of the electromagnetic spectrum.

When dealing with molecular vibrations, it is common practice to use the spectroscopic unit of the wavenumber, cm^{-1}, and in these units, the vibrational frequency of N_2 (or more correctly the vibrational wavenumber) is given as 2558 cm^{-1}. Within the range of wavenumber values defining typical vibrational modes, this places N_2 somewhere towards the higher wavenumber end. More generally, molecular bonds vibrate at frequencies which depend both on the strength of the bond, which is defined by its force constant, and also on the mass being swung about during the course of the vibration according to

$$\nu_i = \frac{1}{2\pi}\sqrt{\frac{k_i}{m_i}}\ s^{-1} \tag{2.21}$$

Here, ν_i is the frequency of a particular vibrational mode, i, k_i is the force constant and m_i is the effective mass of the vibration. Typically then, molecules vibrate at wavenumber values extending from about 200 cm^{-1} at the lower end to as high as 3600 cm^{-1} at the upper end, a range which covers the short to long wavelength region of the infrared.

As we described above, the ways in which molecules vibrate depend fundamentally on the arrangement of the constituent nuclei, on the strengths of interaction which bind them and on the masses of those nuclei. Similarly, the comparability between typical molecular vibrational frequencies and the frequencies of light in the infrared part of the electromagnetic spectrum suggests that the interaction of one with the other might provide a means to elucidate detailed information about molecular structure. Indeed, the action of the electric field component of electromagnetic light with molecular motion involving a change in dipole results in the absorption of that radiation and the excitation of specific modes of vibration.

If we irradiate the molecule with broadband infrared light, there is every prospect that we can excite any number of different vibrational modes and for every mode that is excited we might then expect to see a corresponding reduction in the intensity of light at discrete wavelengths corresponding to the frequencies of vibration of those modes. This is the basis of infrared spectroscopy.

2.11 Infrared spectroscopy

The harmonic oscillator and vibrational energy levels

In treating molecular *rotations* in section 2.3, we started by constructing a simple model of a diatomic molecule as a free rotor in which two nuclei of point masses m_1 and m_2, are separated by a fixed distance, r. In considering molecular *vibrations*, our starting point is to treat each vibrational normal mode of our molecule as an independent harmonic oscillator. Classically, a harmonic oscillator must obey Hooke's Law which states that the force required to compress or extend a spring is proportional to the displacement from equilibrium (see Figure 2.20). This translates to the potential energy varying with the square of the displacement. Thus for a typical bond stretching vibration, the potential energy will vary according to

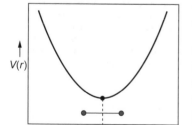

Figure 2.20 Potential energy of a classical harmonic oscillator with r_e the equilibrium bond length.

$$V(x) = \tfrac{1}{2}kx^2 \tag{2.22}$$

where $x = r - r_e$ is the displacement from equilibrium along a particular internal coordinate and k is the force constant.

In quantum mechanical terms, the vibrational energy must take discrete values given by the vibrational term value expression

$$G(v_i) = \omega_i(v + \tfrac{1}{2}) \text{ cm}^{-1} \tag{2.23}$$

where ω_i is known as the vibration wavenumber of a particular normal mode, i, (effectively the vibrational frequency expressed in terms of the spectroscopic unit of cm^{-1})

$$\omega_i = \tilde{v}_i = \frac{v_i}{c} = \frac{1}{2\pi c}\sqrt{\frac{k_i}{m_i}} \text{ cm}^{-1} \tag{2.24}$$

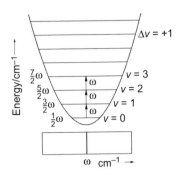

Figure 2.21 The energy levels for a harmonic oscillator are equally spaced.

and m_i is the effective mass of the mode, a measure of the mass being shifted about during the course of the vibration. The vibrational quantum number, v, can take values $v = 0,1,2,3, ...$ with the vibrational term value expression given in Equation (2.23) then yielding an infinite number of equally-spaced energy levels accommodated within a symmetric potential (see Figure 2.21).

We see from Equation (2.23) that adjacent energy levels are separated by ω. The $v = 0$ level, which has an energy of $\tfrac{1}{2}\omega$, is called the zero point level—the minimum vibrational energy the molecule may have, even at absolute zero.

When a molecule absorbs infrared radiation, the molecule is raised from a lower vibrational energy state (typically the zero point level) to a higher vibrational state but that process will only happen if the energy of the photon of light exactly matches the difference in energy between the two quantum states. If this requirement is satisfied AND the vibrational mode concerned involves a change in dipole in the molecule, then the photon is absorbed and the molecule excited.

Within the harmonic approximation, the selection rule restricts transitions to those in which the vibrational quantum number changes by ±1. The implication of this $\Delta v = \pm 1$ selection rule is that for a given mode, all vibrational transitions, regardless of the originating lower level, would yield a spectroscopic line of the same wavenumber, ω (Figure 2.21).

In quantum mechanics, a harmonic oscillator can never be at rest—the nuclei are never completely stationary. Consequently, even at absolute zero, all molecules possess a small amount of vibrational energy known as the zero point energy and the associated energy level is known as the zero point level.

The anharmonic oscillator, fundamentals, and overtones

In reality, a molecular bond vibration only really approximates to harmonic at the bottom of the potential curve, close to the equilibrium separation. The compression of a real chemical bond would result in the potential energy rising more rapidly than suggested by the harmonic potential, whilst for extension, the electronic glue holding the nuclei together will gradually relax its grip and the potential will flatten off, eventually converging to an asymptote at infinite separation corresponding to dissociation of the bond. The resulting potential is then no longer a symmetric *harmonic* potential but an asymmetric *anharmonic* potential (see Figure 2.22).

The vibrational term values for the anharmonic oscillator are given by

$$G(v) = \omega_e(v + \tfrac{1}{2}) - \omega_e x_e(v + \tfrac{1}{2})^2 + \omega_e y_e(v + \tfrac{1}{2})^3 + L \tag{2.25}$$

where ω_e is the vibration wavenumber and $\omega_e x_e$ and $\omega_e y_e$ are anharmonicity constants of rapidly decreasing value. The zero point energy is straightforwardly obtained by substituting $v = 0$ into Equation (2.25) to yield

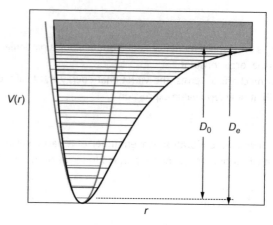

Figure 2.22 For the anharmonic oscillator, the potential energy curve flattens out for large r, resulting in converging vibrational energy levels.

$$G(0) = \tfrac{1}{2}\omega_e - \tfrac{1}{4}\omega_e x_e + \tfrac{1}{8}\omega_e y_e + \cdots$$

(2.26)

In contrast to the harmonic oscillator, the energy levels in this case converge to a dissociation limit as the potential energy flattens out. Above this point, the nuclei are no longer constrained by the potential (they are free) and their energy is then no longer quantized. The dissociation energy D_e is the difference between the energy at the dissociation limit and the equilibrium potential energy but in practice what is measured from experiment will be the zero point dissociation energy, D_0, which is measured relative to the vibrational zero point level.

An important consequence of the change from a harmonic model to an anharmonic model oscillator is that the selection rule modifies to

$$\Delta v = \pm 1, \ \pm 2, \ \pm 3 , \dots$$

(2.27)

For transitions from the zero point level ($v = 0$),

- the $\Delta v = +1$ absorption is known as the *fundamental* absorption;
- the $\Delta v = +2$ is known as the *first overtone* (or second harmonic)
- the $\Delta v = +3$ is known as the *second overtone* (or third harmonic)

with the intensity of the line associated with each successive harmonic diminishing rapidly. In practice, although overtones are allowed and do appear routinely, infrared spectra will be dominated by the much stronger fundamental transitions ($\Delta v = +1$) and it is these lines that are used predominantly to characterize a particular molecular sample. It is worth noting that as the first overtone appears at twice the wavenumber of the fundamental, the second at three times the fundamental wavenumber, and so on, for higher wavenumber fundamentals we quickly find ourselves venturing outside the infrared region and into the visible (see Figure 2.23).

So far, we have seen that the absorption of infrared light by a molecule will lift it from, typically, its zero point vibrational level to some excited vibrational level, with the most likely outcome being one where the vibrational quantum number changes by 1 unit. In a diatomic molecule, there is only one vibrational degree of freedom corresponding to a single bond stretching mode but we must also remember that

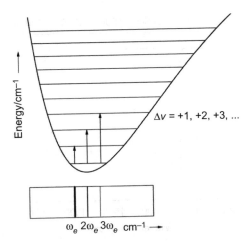

Figure 2.23 For the anharmonic oscillator, overtones are allowed but with rapidly diminishing intensity.

light can only be absorbed if the resulting vibration causes a change in dipole. Consequently, a *homonuclear* diatomic molecule, being non-polar, will, by definition, not display an infrared spectrum but a *polar heteronuclear* diatomic molecule will. However, this does not mean that all non-polar molecules fail to absorb infrared light. In our discussion earlier, we stated that a requirement for infrared absorption is that excitation of a particular vibrational mode must lead to a change in dipole in the molecule. Thus a non-polar linear molecule such as CO_2 will display an infrared spectrum but only when vibrational modes are excited which result in the creation of a dipole. Such modes are referred to as *infrared active*.

In CO_2, there are 4 vibrational modes: the *symmetric stretch*, in which both C–O bonds extend and compress in phase with one another; the *anti-symmetric stretch* in which one C–O bond extends whilst the other compresses; and a pair of so-called degenerate *bending modes* in which the O–C–O bond angle varies (there are two of these because the bond can bend within two mutually perpendicular planes).

It should be clear from Figure 2.24 that exciting the symmetric stretch simply renders the molecule longer or shorter about the internuclear axis, with no change to its polarity. However, excitation of the anti-symmetric stretch or the bend creates an asymmetry in the charge distribution in the molecule thereby creating a dipole which then varies in magnitude throughout the course of the vibration. Thus both the anti-symmetric stretch and the pair of bends are *infrared active* modes whilst the symmetric stretch is *infrared inactive*.

The infrared spectrum of CO_2 shown in Figure 2.25 exhibits just two bands: the intense band at about 2364 cm^{-1} is due to the fundamental ($v = 0 \rightarrow v = 1$) in the anti-symmetric stretching mode with the strong intensity reflecting the large induced dipole change that accompanies the vibration; the much weaker band at about 670 cm^{-1} is due to the bending mode and its weak intensity reflects the much smaller induced dipole change that accompanies this vibration. As the symmetric stretch involves no change in dipole, it is infrared inactive and does not appear at all in the spectrum. The high wavenumber of the stretch and relatively low wavenumber of the bend reflects the degree of floppiness of each vibration. It is generally the case that stretches are

In the context of molecular energy levels, degeneracy refers to two or more states of the system having identical energy. In the case of molecular vibrations, such as in the two mutually perpendicular bending vibrations in CO_2, it should be apparent that the natural bending frequency will not depend on the direction from which the vibrating molecule is being observed.

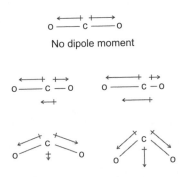

Figure 2.24 Normal modes which result in changing dipole moment are IR active.

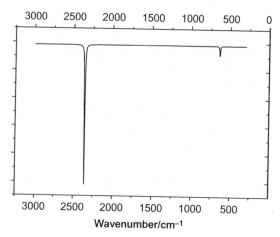

Figure 2.25 The infrared spectrum of CO_2 exhibits just two bands. The higher wavenumber band is due to the anti-symmetric stretching vibration whilst the lower wavenumber band is due to the degenerate pair of bending modes.

of much higher wavenumber (or frequency) compared with bends where the same nuclei are involved.

2.12 Vibrational Raman spectroscopy

The Rayleigh line in a vibrational Raman spectrum is generally only used as a zero wavenumber point of reference relative to the positions of the vibrational lines appearing to lower wavenumber. However, at higher resolution we see that the Rayleigh line is composed of a sharp central feature with a series of discrete lines either side. These features are simply the rotational Raman spectrum shown in Figure 2.16 and discussed in detail in section 2.8.

The spectroscopic process described above is an example of classical optical absorption in which the energy difference between discrete vibrational quantum states is bridged by infrared radiation of the correct frequency. However, as with rotational Raman described earlier in section 2.8, changes in *vibrational* state can also be achieved by Raman scattering, with the incident photon either losing energy (Stokes scattering) or gaining energy (anti-Stokes scattering). The only real difference in vibrational Raman is that the inelastically scattered photons will have lost or gained more energy as a result of the interaction than is the case in rotational Raman spectroscopy because the energy gaps between the vibrational energy levels are much larger than between rotational levels. The larger energy level spacing also means that the anti-Stokes lines will be much less intense than in rotational Raman because anti-Stokes scattering can only occur from vibrationally excited levels and these will be populated to a much smaller extent than rotationally excited levels for a given temperature (see Figure 2.26). Consequently, a vibrational Raman spectrum typically only presents the Stokes lines, referenced to the very intense Rayleigh line, with the anti-Stokes component ignored.

A vibrational Stokes or anti-Stokes transition will occur for any vibrational mode which results in a variation in the *polarizability* of the molecule. Such modes are known as *Raman active*, just as modes which result in a changing dipole are known as *infrared active*. In spite of the different gross requirements between the two techniques for a mode to be active, the $\Delta v = \pm 1, \pm 2, \pm 3, \ldots$ selection rule is the same in both cases. Thus, just as with infrared spectroscopy, we can see fundamental transitions as well as higher overtones (see Figure 2.27).

For molecules of higher symmetry, some of the modes which are Raman active may be infrared inactive and some of the infrared active modes may be Raman inactive. Such differences in IR and Raman activity of different modes means that the two

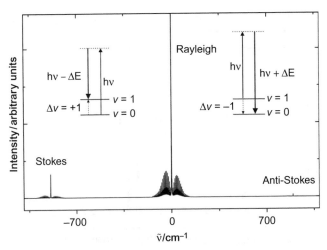

Figure 2.26 The vibrational Raman spectrum of F_2. The intense Rayleigh line at 0 cm^{-1} corresponds to elastically scattered light having neither gained nor lost energy. The bands immediately adjacent to the Rayleigh line on either side are the rotational Raman Stokes and anti-Stokes lines (see Figure 2.16) whilst the bands at -893 cm^{-1} and $+893$ cm^{-1} are the vibrational Raman Stokes and anti-Stokes lines, respectively.

techniques often provide far more insight into molecular structure as a pair than either one might play in isolation. It is this complementarity which imparts such tremendous power to use of these two methods in molecular structure determination.

If we return to the example of CO_2, it turns out that none of the infrared active modes generate a change in the polarizability of the molecule during the vibration but the one infrared inactive mode, the symmetric stretch, does. Consequently, the Raman spectrum of CO_2 should display just a single band associated with the symmetric stretch, with the two bands which appear in the infrared spectrum failing to appear in the Raman spectrum. This *mutual exclusivity* in the IR and Raman activity in CO_2 derives from the fact that it possesses what is known as a *centre of inversion* (see Figure 2.28).

A centre of inversion is a symmetry property that exists in molecules possessing both a two-fold rotational symmetry (i.e. where a 180° rotation about an axis leaves the molecule indistinguishable from its starting point) AND a plane of reflection perpendicular to that rotational axis. All such molecules are subject to what is known as the *Mutual Exclusion Principle* which states that

'for a molecule with a centre of inversion, the fundamentals which are active in the Raman spectrum are inactive in the infrared spectrum whilst those active in the infrared spectrum are inactive in the Raman spectrum'.

Consequently, where a pair of infrared and Raman spectra display mutual exclusivity in the location of the vibrational bands appearing in each, it is reasonable to conclude that the molecule concerned possesses an inversion centre.

A quirk of CO_2 is that although it possesses a centre of inversion and should therefore provide an excellent example of the mutual exclusion principle, the Raman spectrum contains two bands where we might reasonably only expect to see one (see Figure 2.29). It just so happens that the energy level corresponding to a single quantum

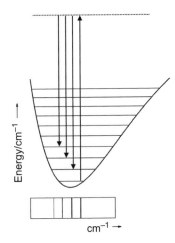

Figure 2.27 Fundamental, first and second overtone transitions resulting from vibrational Raman scattering.

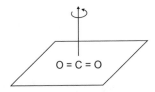

Figure 2.28 A centre-of-inversion is a symmetry property of a molecule possessing both a two-fold rotational symmetry AND a plane of reflection perpendicular to that axis of rotation.

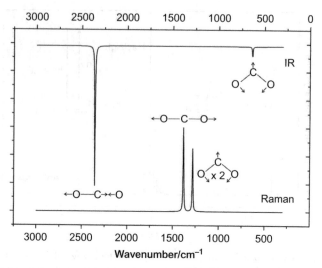

Figure 2.29 The infrared active modes in CO_2 are Raman inactive whilst the Raman active modes are infrared inactive. This mutual exclusivity is a characteristic of molecules possessing a centre of inversion (see Figure 2.28). Thus the antisymmetric stretch and bending fundamentals appear only in the infrared spectrum while the symmetric stretch appears only in the Raman spectrum. A complicating factor in the case of CO_2 is that the symmetric stretch appears in the Raman spectrum as a Fermi resonance with the first overtone in the bending mode (see text).

change in the symmetric stretch (≈ 1350 cm^{-1}) lies at almost exactly the same energy as two quanta in the bend (2×670 cm$^{-1} = 1340$ cm^{-1}) and as a result the two levels resonate together in a so-called Fermi resonance yielding a pair of bands close together rather than just a single band due to the symmetric stretch fundamental. The first overtone in the bend is actually Raman allowed because although excitation of a single quantum of bend does not yield a changing polarizability, excitation of two quanta does! Normally, in spite of the fact that the bending overtone is Raman active, we might expect the intensity to be weak but the Fermi resonance results in the bend overtone borrowing intensity from the strongly allowed stretching mode and the two bands appear with comparable intensity in the Raman spectrum.

2.13 Vibration–rotation spectroscopy

Infrared and Raman spectroscopy are perhaps most widely used in the routine characterization of samples in the condensed phase and as such are analysed in terms of the association between vibrational bands and vibrational modes (see section 2.14). However, in the gas phase, molecules have a great deal more freedom to exercise their full repertoire of motion and, as we saw in the first part of this chapter, that includes rotational motion. For example, the spectrum shown in Figure 2.30 is an infrared spectrum taken from the output of a car exhaust and shows vibrational bands attributed to water, CO, CO_2, methane and unspent hydrocarbons. A particularly striking feature of this spectrum is that rather than the amorphous, smooth bands that appear in a liquid phase infrared spectrum, all of the bands exhibit an underlying fine structure and in the case of the water and CO bands, that structure resolves into a multitude of individual lines. This additional fine structure, which we also saw in the Raman spectrum of F_2 shown in Figure 2.26, is of course *rotational* fine structure.

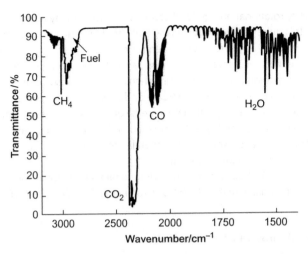

Figure 2.30 Infrared spectrum of the mixture of gases in a car exhaust.

Each vibrational level has associated with it a stack of rotational levels (see Figure 2.31). In sections 2.4 and 2.8 we saw that we can observe transitions between *rotational levels* associated with the same vibrational level through either absorption or emission of light in the microwave and millimetre wave parts of the electromagnetic spectrum or, in rotational Raman spectroscopy, through the inelastic scattering of (typically) visible light. In *vibrational* spectroscopy, we observe transitions between different vibrational levels through absorption of infrared radiation (or, again through the Raman process) but, when we do so in the gas phase, we will simultaneously be exciting transitions between stacks of discrete rotational levels associated with each of the different vibrational levels. As a consequence, we may see rotational fine structure bands associated with each vibrational band.

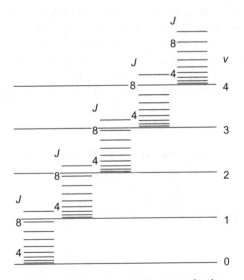

Figure 2.31 Vibration–rotation energy levels.

In principle, rotational energy level changes will accompany all such vibrational transitions. It is quite easy to imagine why that should be the case: we saw in Equations (2.1) and (2.2) that the rotational energy is related to the moment of inertia and that the moment of inertia is related to the square of the bond length. The instantaneous change in vibrational state that occurs when a molecule absorbs or emits infrared light will result in an instantaneous increase or decrease in the bond length and this will necessarily result in a proportionate increase or decrease in the rotational energy. However, rotational motion is greatly impeded in the condensed phase and consequently vibration–rotation spectra can generally only be observed in the gas phase and then preferentially at lower pressures.

Remember that *term values* is the term used when describing energy levels presented in terms of the spectroscopic unit of cm^{-1}.

The term values, T, describing the total vibrational and rotational energy are given as a sum of vibrational (Equation (2.25)) and rotational (Equation (2.4)) term values:

$$T = G(v) + F_v(J)$$
$$= \omega_e(v + \tfrac{1}{2}) - \omega_e x_e(v + \tfrac{1}{2})^2 + \cdots + B_v J(J+1) \tag{2.28}$$

For most diatomic molecules and some linear polyatomic molecules, the rotational selection rules governing transitions between two stacks of levels is $\Delta J = \pm 1$, yielding an R branch ($\Delta J = +1$) and a P branch ($\Delta J = -1$). Each transition is labelled $R(J)$ or $P(J)$ where J represents the J-value of the lower state, J''. A consequence of the $\Delta J = \pm 1$ selection rule is that the pure vibrational transition, for which $\Delta J = 0$, is not observed (see Figure 2.32). However, the position at which it would appear is known as the band centre and is labelled ω_0.

As with any form of optical spectroscopy, the lines that appear in the spectrum are the result of transitions between energy levels in accordance with the selection rules and consequently, their energy positions in the resulting spectrum are obtained from the *differences* between the energies of the levels involved in the transition. One of our opening justifications in this section was to reason why rotational energy change should accompany vibrational energy change and it is similarly reasonable to suggest that the rotational constant associated with an excited vibrational level might be different from that associated with the lower lying level. However, in considering how the rotational fine structure might arrange itself in a vibration–rotation spectrum involving a $v = 1-0$ transition (the fundamental transition from $v = 0$ to $v = 1$), a sensible initial assumption is that the rotational constant associated with the upper vibrational state, B_1, is approximately the same as that associated with the vibrational zero point level, B_0, and so we assume for the purpose of our initial treatment that

$$B_1 \approx B_0 = B$$

If we do this, then the wavenumber positions of the lines in the R branch of the vibration–rotation spectrum for which $\Delta J = +1$ will be

$$\bar{v}(R(J)) = \omega_0 + B(J+1)(J+2) - BJ(J+1)$$
$$= \omega_0 + 2BJ + 2B \tag{2.29}$$

and for the P branch

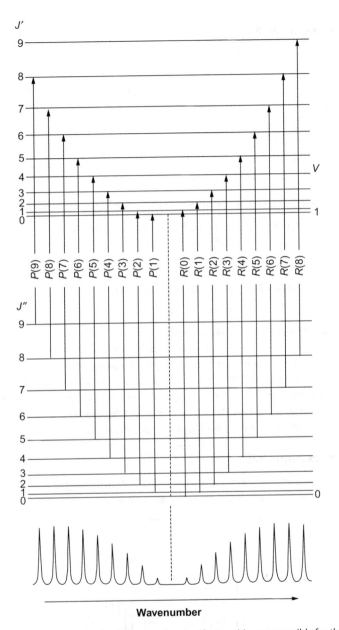

Figure 2.32 Schematic energy level diagram showing the transitions responsible for the fine structure in a vibration–rotation band for a diatomic molecule.

$$\tilde{v}(P(J)) = \omega_0 + BJ(J-1) - BJ(J+1)$$
$$= \omega_0 - 2BJ \qquad (2.30)$$

where ω_0 is the wavenumber position of the band centre corresponding to the pure vibrational transition (for which $\Delta J = 0$).

Worked example 2.5

Question. The zero band gap is the conspicuous gap in the centre of the vibration-rotation band between the first member of the R branch, $R(0)$, and the first member of the P branch, $P(1)$ (see Figure 2.32). Use Equations (2.29) and (2.30) to show the zero band gap, $\tilde{v}(R(0)) - \tilde{v}(P(1))$, is $4B$ and that the spacing between adjacent lines in either the P or R branches is $2B$.

Answer. The zero band gap is given by

$$\tilde{v}(R(0)) - \tilde{v}(P(1)) = \omega_0 + (2B \times 0) + 2B - (\omega_0 - 2B \times 1)$$
$$= 2B - (-2B) = 4B \tag{2.31}$$

The separation between any two adjacent lines in the R branch is given by

$$\tilde{v}(R(J+1)) - \tilde{v}(R(J)) = \omega_0 + 2B(J+1) + 2B - \omega_0 - 2BJ - 2B$$
$$= 2B(J + 1 - J) = 2B \tag{2.32}$$

and in the P branch

$$\tilde{v}(P(J)) - \tilde{v}(P(J+1)) = \omega_0 - 2BJ - \omega_0 + 2B(J+1)$$
$$= 2B(-J + J + 1) = 2B \tag{2.33}$$

It should be clear that a vibration-rotation spectrum can be used in much the same way as a pure rotation spectrum to obtain a value for the rotational constant, B, and hence the moment of inertia and bond length for a diatomic molecule. In using any one of Equations (2.31) to (2.33) to do so, we are assuming that the rotational constant in the upper vibrational state is the same as that in the lower state and consequently we are likely to obtain a bond length similar to, but not exactly the same as, the zero point bond length, r_0, obtained from a pure rotational spectrum.

Worked example 2.6

The infrared vibration-rotation spectrum of CO is shown in Figure 2.33 (see also the infrared spectrum of car exhaust gases in Figure 2.30).

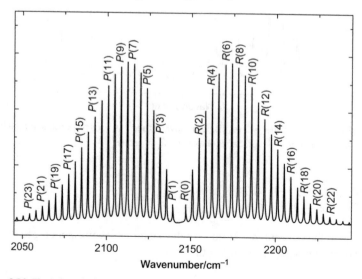

Figure 2.33 The infrared vibration-rotation spectrum of CO.

The wavenumber positions of the first ten lines either side of the band centre are given in Table 2.2.

Question.

(a) Use the data in Table 2.2 to find the wavenumber value of the pure vibrational transition and hence calculate the force constant, k, for the CO bond.

(b) Calculate the average separation of adjacent lines in the R and P branches and hence find the rotational constant, B.

(c) Use your answer from part (b) to calculate the CO bond length.

Answer.

(a) The wavenumber of the pure vibrational transition will lie approximately half way between the $R(0)$ and $P(1)$ lines. Thus with $R(0) = 2147.0$ cm^{-1} and $P(1) = 2139.4$ cm^{-1}, we find that $\omega_0 = (2147.0 + 2139.4)/2 = 2143.2$ cm^{-1}.

We know from Equation (2.24) that the vibration wavenumber is related to the force constant, k, according to

$$\omega_0 = \frac{1}{2\pi c}\sqrt{\frac{k}{\mu}}\ \text{cm}^{-1} \tag{2.34}$$

Here we have replaced the effective mass, m_i, for a particular mode, i, in a polyatomic molecule, with the reduced mass, μ, for a diatomic molecule, where in this case

$$\mu_{CO} = \frac{m_C m_O}{m_C + m_O}$$

The absolute reduced mass of CO was calculated in Worked example 2.1 as 1.139×10^{-26} kg and so we can now simply rearrange Equation (2.34) to find the force constant, k:

$$k = \mu_{CO}(2\pi c\omega_0)^2 = 1855.2\ \text{N m}^{-1}.$$

(b) Using the appropriate Δ values shown in the table, the eighteen pairs of adjacent lines in the R and P branches are separated by an average value of 3.83 cm^{-1}. We know from Worked example 2.5 that pairs of adjacent lines are separated by $2B$ and so it follows that the rotational constant $B = 1.915$ cm^{-1}.

(c) Following the same procedure as in Worked example 2.1, we find the bond length of CO as

$$r_{CO} = \sqrt{\frac{6.626 \times 10^{-34}\ \text{J s}}{8\pi^2 \times 1.915\ \text{cm}^{-1} \times 1.139 \times 10^{-26}\ \text{kg} \times 2.997 \times 10^{10}\ \text{cm s}^{-1}}}$$

$$= 1.133 \times 10^{-10}\ \text{m} = 1.133\ \text{Å}$$

You will notice that this value, which is derived from data from both the $v=0$ and $v=1$ vibrational levels, is slightly larger than the value of 1.132 Å we obtained from the pure rotational spectrum. This observation is in keeping with our expectation that the average bond length should be larger in a vibrationally excited level than in the zero point level.

So how exactly does a vibration–rotation infrared spectrum provide additional value over a pure rotational spectrum in terms of molecular structure determination? Well, we can see from the table in Worked example 2.6 that the separations between adjacent lines in the P branch are larger than those in the R branch—in other words, the spectrum shown in Figure 2.33 is not quite as symmetrical as it first appears. The

Table 2.2

Line	Wavenumber/ cm^{-1}	Δ/cm^{-1}
$R(9)$	2179.8	3.5
$R(8)$	2176.3	3.5
$R(7)$	2172.8	3.6
$R(6)$	2169.2	3.6
$R(5)$	2165.6	3.6
$R(4)$	2162.0	3.7
$R(3)$	2158.3	3.7
$R(2)$	2154.6	3.8
$R(1)$	2150.8	3.8
$R(0)$	2147.0	7.6
$P(1)$	2139.4	3.9
$P(2)$	2135.5	3.9
$P(3)$	2131.6	4.1
$P(4)$	2127.5	3.8
$P(5)$	2123.7	4.0
$P(6)$	2119.7	4.0
$P(7)$	2115.7	4.1
$P(8)$	2111.6	4.2
$P(9)$	2107.4	4.1
$P(10)$	2103.3	-

Note: This is experimental data and as such will be subject to the usual uncertainties associated with experimental error.

The pure vibration wavenumber ω_0 corresponds to the difference in energy between the $v = 0$, $J'' = 0$ level and the $v = 1$, $J' = 0$ level. If our molecule were a harmonic oscillator, then this quantity gives the absolute vibrational wavenumber of the molecule which is directly analogous to the classical vibrational frequency of a harmonic oscillator. In reality, CO is not a harmonic oscillator but an anharmonic oscillator and so we need to keep in the back of our minds that ω_0 is not quite the same as ω_e given in Equation (2.25). For the purposes of this discussion though, we can treat the two as approximately equivalent. A vibrational wavenumber of 2143 cm^{-1} for CO corresponds to a classical vibrational frequency of 6.42×10^{13} Hz. That is, sixty four trillion vibrations per second.

reason for this lies in the fact that the rotational constants, B_0 and B_1, associated with the two vibrational levels are not quite the same, as we had assumed in deriving Equations (2.29) and (2.30). A more rigorous treatment of the data should allow us therefore to estimate the values of B_0 and B_1 separately and then to use that information to find B_e, the rotational constant associated with the hypothetical equilibrium state of the molecule at the bottom of the potential well. We should then be in a position to find the equilibrium bond length, r_e.

Variations on a theme: the differences between r_e, r_0, and r_1

In presenting rotational and vibration–rotation spectroscopy as a means to measure molecular bond lengths, we have rather unfortunately thrown up the inconvenient truth that what you get from the measurement of a bond length really rather depends on which experiment you choose to perform the measurement. In getting this far, we have obtained a measurement of r_0, the average bond length of CO in its zero point level from a pure rotational spectrum. We have also obtained a bond length averaged over two different vibrational states in CO from the vibration–rotation spectrum where we have assumed that the rotational constant is the same in both of the two vibrational states. And we are already familiar with the notion of the equilibrium bond length, r_e, but may not have quite appreciated that in the quantum world this is a purely hypothetical concept. Nevertheless, it is possible to obtain a measurement for r_e, but in doing so we have to exploit the fact that the rotational constant does vary with vibrational state and that each vibrational state will yield its own bond length, $r_1, r_2, r_3, \ldots$ from measurements of rotational constants associated with each vibrational level. We then use two or more of those to extrapolate back to a value for r_e. To illustrate the process, let's limit ourselves to the first two vibrational levels and come up with an approach that will allow us to obtain measurements of the two rotational constants, B_0 and B_1.

The method we shall use to find B_0 and B_1 is known as the method of combination differences which uses as its basis the observation that the differences in wavenumbers for transitions with a common upper state will be dependent only on the properties of the lower state while differences in wavenumbers for transitions with a common lower state will be dependent only on the properties of the upper state. So, for example, we can see from Figure 2.32 that the R(0) and P(2) transitions both have in common the $J' = 1$ level whilst the R(2) and P(2) transitions both have the $J'' = 2$ level in common. In general, differences in transitions with a common upper state J-level will depend only on the lower state rotational constant, B'', according to

$$\Delta F''(J) = \tilde{v}(R(J-1)) - \tilde{v}(P(J+1))$$
$$= 4B''(J+\tfrac{1}{2}) \tag{2.35}$$

where in this example, $B'' = B_0$. Similarly, differences in transitions with a common lower state J-level will depend only on the upper state rotational constant, B', according to

$$\Delta F'(J) = \tilde{v}(R(J)) - \tilde{v}(P(J+1))$$
$$= 4B'(J+\tfrac{1}{2}) \tag{2.36}$$

where $B' = B_1$.

Worked example 2.7

Question. Use Equations (2.35) and (2.36) to find B_0 and B_1 for CO and hence determine the average bond lengths r_0 and r_1, in each of the two vibrational states.

Answer. A plot of $\Delta F''(J)$ against $(J + \frac{1}{2})$ should yield a straight line with a gradient of $4B_0$. The transitions with a common upper level are $R(0)$ and $P(2)$, $R(1)$ and $P(3)$, $R(2)$ and $P(4)$ and so on. Their differences are shown in Table 2.3.

A plot of $\Delta F''(J)$ against $(J + \frac{1}{2})$ yields a straight line:

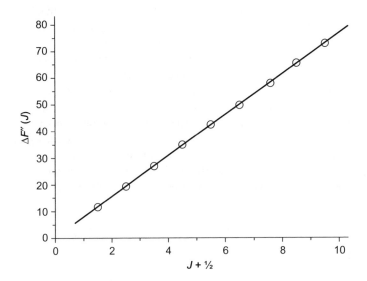

Table 2.3

J'	$\Delta F''(J)/\text{cm}^{-1}$
1	$R(0) - P(2) = 11.5$
2	$R(1) - P(3) = 19.2$
3	$R(2) - P(4) = 27.1$
4	$R(3) - P(5) = 34.6$
5	$R(4) - P(6) = 42.3$
6	$R(5) - P(7) = 49.9$
7	$R(6) - P(8) = 57.6$
8	$R(7) - P(9) = 65.4$
9	$R(8) - P(10) = 73.0$

The gradient is $7.655 = 4B_0$ from which we obtain a value for the zero point level rotational constant $B_0 = 1.914 \text{ cm}^{-1}$. This yields a value for r_0 of 1.133 Å which is in very good agreement with the value of 1.132 Å obtained in Worked example 2.1 from the pure rotational spectrum.

The transitions with a common lower level are $R(1)$ and $P(1)$, $R(2)$ and $P(2)$, $R(3)$ and $P(3)$ and so on. Their differences are presented in Table 2.4.

Plotting $\Delta F'(J)$ against $(J + \frac{1}{2})$ for this series of transitions yields a value for the rotational constant $B_1 = 1.896 \text{ cm}^{-1}$ which gives a value for r_1 of 1.139 Å.

What is clear from the results of Worked example 2.7 is that in CO there is a small but significant dependence of rotational constant on vibrational level. As the level of vibrational excitation increases so does the effective average bond length of the molecule and the longer bond length translates to a larger moment of inertia and a smaller rotational constant.

The vibrational dependence of B is given by

$$B_v = B_e - \alpha_e(v + \tfrac{1}{2}) \tag{2.37}$$

where B_e is the rotational constant at the hypothetical equilibrium bond length and α_e is the vibration–rotation interaction constant. A plot of B_v against $v + \frac{1}{2}$ using the two data points we have yields a gradient $\alpha_e = 0.018 \text{ cm}^{-1}$ and a y-axis intercept $B_e = 1.923 \text{ cm}^{-1}$. We are now in a position, finally, to obtain a value for the equilibrium bond length, r_e, for CO of 1.131 Å.

Table 2.4

J''	$\Delta F'(J)$
1	$R(1) - P(1) = 11.4$
2	$R(2) - P(2) = 19.1$
3	$R(3) - P(3) = 26.7$
4	$R(4) - P(4) = 34.5$
5	$R(5) - P(5) = 41.9$
6	$R(6) - P(6) = 49.5$
7	$R(7) - P(7) = 57.1$
8	$R(8) - P(8) = 64.7$
9	$R(9) - P(9) = 72.4$

In diffraction experiments (electron, X-ray, and neutron), the experimental bond lengths are reported as mean bond lengths in recognition of the vibrational averaging implicit in the acquisition of the diffraction pattern over relatively long molecular time scales. In this respect, average bond lengths reported from diffraction data are not quite the same as the effective zero point bond length obtained from spectroscopic measurements. Nevertheless, equilibrium bond lengths can be obtained from diffraction data by reduction of the mean bond lengths through analysis of the vibrational coordinate in terms of a model anharmonic potential.

This exercise has provided a certain amount of food for thought from two perspectives. On the one hand, we seem to have gone to an awful lot of trouble to find three different values for a molecular bond length, none of which seems very different from the value we obtained from the pure rotational spectrum in Worked example 2.1. So whether you go to all of this trouble really rather depends on the level of precision you demand in your determination of this particular molecular structure parameter. The second perspective of course derives from a fuller consideration of what exactly we mean when we talk about bond length. We get into an easy habit of dealing with equilibrium bond lengths but the fact remains that in the quantum mechanical world, the equilibrium bond length is a hypothetical concept because the molecule can never reside at the bottom of the potential well. Instead, at any given temperature, the majority of molecules will sit aloof from that classical conceit in their zero point levels or higher, and consequently, a more meaningful value for the bond length is r_0. The fact that $r_0 > r_e$ reflects the fact that the most probable separation of the two nuclei when the molecule has a non-zero amount of vibrational energy (which it does even at absolute zero) will be one which reflects the fact that the bond is undergoing a degree of vibrational motion, extending and compressing, but because of the shape of the anharmonic potential, tending to extend slightly more than it compresses. This tendency is only exaggerated further as the molecule is lifted into higher vibrational levels which is why we found $r_1 > r_0$ in the case of CO.

2.14 Group vibrations, chemical characterization, and analysis

Infrared spectra get very rapidly more complicated as we progress from diatomic molecules to triatomic and then on to larger polyatomic molecules because the number of vibrational modes increases in proportion to the number of nuclei (remember there are 3N–6 vibrational modes for a non-linear molecule, and 3N–5 for a linear molecule). So, for a molecule as apparently simple as fluorobenzene, there will be 30 vibrational modes, 27 of which are infrared active, and which may in principle appear in an infrared spectrum as fundamentals, overtones but also in combination with other modes as combinations (e.g. $v_2 + v_3$) or as differences (e.g. $v_2 - v_3$). As the number of vibrations increases, it becomes increasingly difficult to represent each mode in terms of a simple internal coordinate description particularly for large molecules with low symmetry, and consequently the vibrational spectrum becomes increasingly difficult to interpret. In general, a full analysis and assignment of a spectrum is really only possible for small, simple molecules or those of particularly high symmetry but nevertheless it is often possible to obtain useful information about the structure of complex molecules because their spectra exhibit characteristics which can be associated with particular structural features, which when combined with other characteristic bands, allow a clear association between the spectrum and a particular molecule.

Although vibrational spectra often comprise numerous absorption bands whose assignment may be problematic, they also frequently exhibit bands which may be associated with particular functional groups. In general, normal mode vibrations will involve motion of all of the nuclei in the molecule but sometimes, the structure of the molecule is such that some of the vibrations localize in particular parts of the

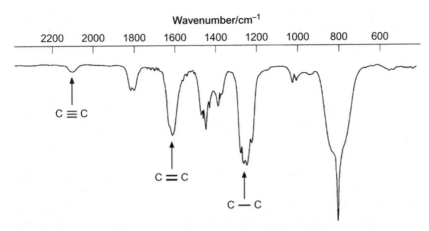

Figure 2.34 The infrared spectrum of 2-methyl-1-butene-3-yne exhibits bands at 1250, 1620, and 2120 cm^{-1} which are assigned to alkane, alkene and alkyne carbon-carbon stretching vibrations. The quite different force constants associated with the single, double and triple bonds means that excitation of a vibration in any one of these bonds is largely decoupled from the rest of the molecule with the bonds then vibrating independent of one another.

molecule. Such modes are known as group vibrations and occur typically in one of two situations:

- When the force constant between two atoms in a chain is very different from those between other atoms in the chain.
- When a large difference in mass exists between the nuclei of a terminal functional group.

A good example of the former is given by the molecule 2-methyl-1-butene-3-yne in which a C–C single bond sits between an alkyne (C≡C) group and an alkene (C=C). The force constants for the three groups are sufficiently different to one another that the C–C, C=C and C≡C vibrations can be excited largely independent of one another, with bands associated with vibration of each bond appearing at quite different vibration wavenumbers in the infrared spectrum (see Figure 2.34).

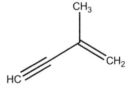

In the case of large differences in mass between the nuclei in terminal groups, good examples are provided by the O–H stretch in ethanol or the C–H stretching modes in the methyl groups of propanone (see Figure 2.36). In either case, the hydrogen atom is so light compared to the much heavier oxygen or carbon that the heavier nucleus is hardly perturbed at all by the furiously vibrating hydrogen and effectively behaves similarly to a wall reacting to the impact of a tennis ball. As the vibrational frequency or wavenumber is inversely proportional to the square root of the effective mass of the vibration, where in this case the effective mass is almost entirely defined by the mass of the hydrogen atom, we would expect both OH and CH stretching modes to be of high wavenumber. Indeed, typically OH stretches are observed between 3590 and 3650 cm^{-1} with CH stretches a little lower in wavenumber appearing between about 2950 and 3300 cm^{-1}, depending on the group. While the positions of bands associated with group vibrations tend to be consistent from molecule to molecule, so does the band intensity which depends on the magnitude of the dipole being exercised by the vibration. The C≡C stretching mode in 2-methyl-1-butene-3-yne appears quite weakly in the infrared spectrum in Figure 2.34 because the dipole moment is weak but the stretching vibration of a polar OH or C=O group will yield a much stronger absorption band.

The *functional group* region covers about 1500–3700 cm^{-1}, with lower wavenumbers often associated with modes involving strong coupling between stretches and bends in straight or branched chains and rings. The patterns of bands that appear in this region tend to be very specific to particular molecules and for this reason this lower wavenumber region is commonly known as the fingerprint region.

The consistency in position and intensity of bands associated with particular functional groups has meant that vibrational spectroscopy, and infrared spectroscopy in particular, has been used very successfully in characterization of molecular samples. In such applications, the compound under investigation is examined, where possible, as a liquid sample. This can be done, for example, for a neat liquid by squeezing a few drops between two KBr discs (transparent to infrared radiation). Alternatively, the compound may be dissolved in a suitable solvent (e.g. $CHCl_3$, CCl_4), and, although bands characteristic of the solvent will then also appear in the spectrum, these can easily be recognized and allowance made for them. In fact, in double-beam spectrometers, a separate cuvette containing the solvent only is sampled by the second reference beam, and its spectrum may then be subtracted automatically from that of the sample under investigation. If the sample is a solid compound, it may be ground up to form a 'mull' with paraffin oil before being placed between the KBr discs or ground with KBr and pressed into a disc. In the liquid phase, rotational fine structure is smeared out because rotational motion is greatly impeded and so the bands that appear in the spectrum are more straightforwardly just associated with excitation of particular vibrational mode fundamentals, overtones, combinations, and differences.

Clues about an unknown compound's molecular structure that are provided by an infrared spectrum derive from the typical absorptions seen in different parts of the spectrum (see Figure 2.35 for vibrations typically associated with organic molecules). As we have seen previously, the wavenumber value of a vibration will scale with

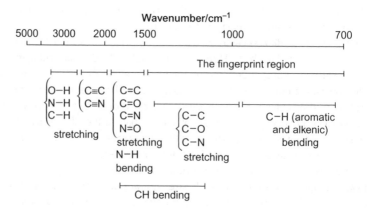

Figure 2.35 Typical absorptions seen in different parts of the infrared spectra of organic molecules.

increasing force constant for the bond involved and with decreasing effective mass (or reduced mass when considering the motion of just two interconnected nuclei). For example, the lowering of the wavenumber (and hence energy) for stretching in the series *triple bond > double bond > single bond* correlates straightforwardly with the appropriate bond enthalpies (bond strengths) and hence force constants. Similarly, we associate high wavenumber (and hence frequency) with C–H, N–H, and O–H stretches (for which the reduced mass is small) but would expect a C–Br stretch to appear to relatively low wavenumber (for which the reduced mass is quite large). It is also generally the case that the bending vibrations are of lower energy than stretches and consequently appear at lower wavenumbers than stretching modes involving the same bonds.

One of the particular attractions of infrared spectroscopy is that, in spite of the premise that group vibrations in their consistency provide a route to deducing structural features in molecular samples, analysis of the positions of infrared bands associated with particular group vibrations, such as the C–H stretch, in a variety of different molecules shows that the exact wavenumber does in fact depend to a small but significant extent on the environment of that bond in a molecule. There are obvious analogies to be made in this respect with the notion of chemical shift in NMR spectra—see Chapter 4) and it is this sensitivity to chemical environment which endows the method with additional diagnostic power. Some characteristic absorptions which show such variation are listed in Table 2.5 and these are now discussed in more detail.

(i) C–H bonds. Infrared absorption bands associated with C–H stretching modes in methyl (CH_3) and methene (CH_2) groups usually appear in the range 2850–2950 cm^{-1}. In the case of the former, there will be three independent CH stretching modes and in the latter case, two. These independent vibrations may be of sufficiently different wavenumber to appear as distinct bands in their own right or close enough together that bands appear with apparent splittings or asymmetry in their shape. The C–H bending modes in both groups appear at around 1450 cm^{-1}. The absorptions for the C–H stretch in propanone at just below 3000 cm^{-1} and the bending modes at *ca.* 1400 cm^{-1} are clearly visible in the infrared spectrum shown in Figure 2.36.

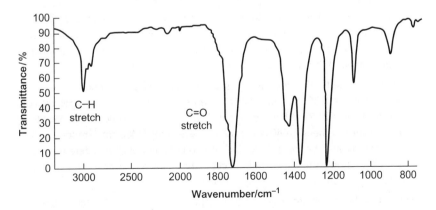

Figure 2.36 Infrared spectrum of propanone, $(CH_3)_2CO$ (liquid film).

Table 2.5 Characteristic infrared absorptions for a variety of organic molecules*

Molecule or group	Vibration type	Wavenumber/cm^{-1}
Alkyl group (CH_3, CH_2, CH)	C–H stretch	2850–2960
	C–H bend	1370–1460
Alkanal (CHO)	C–H stretch	2700–2900
Alkyne (C≡CH)	C–H stretch	3270–3300
Alkene (C=CH_2)	C–H stretch	3075–3095
	C–H bend	890–990†
Arene	C–H stretch	3010–3040
	C–H bend: in-plane	1000–1300
	C–H bend: out-of-plane	650–900†
Alkanol (OH)	O–H stretch	3590–3650$^\Delta$
	C–O stretch	1050–1200
Amine, amide (NH_2)	N–H stretch	3300–3500$^\Delta$
Aliphatic ketone (R_2CO)	C=O stretch	1700–1740
Aliphatic alkanal (RCHO)	C=O stretch	1720–1740
Aromatic ketone (Ar_2CO)	C=O stretch	1680–1700
Alkanoic acid (RCO_2H)	C=O stretch	1700–1725
Alkanoyl chloride (RCOCl)	C=O stretch	1790–1815
Alkanoate ester (RCO_2R')	C=O stretch	1730–1750
	C–O stretch	1050–1300
Alkoxy (ether) R_2O	C–O stretch	1070–1150

* Some of these bands may be split into several components.
† Characteristic variations occur with different substitution patterns.
$^\Delta$ These may be drastically affected by hydrogen-bonding.

Alkanal C–H stretches can appear as low as 2700 cm^{-1} and as high as 2900 cm^{-1}. The C–H stretch in alkynes, C≡CH, (3300 cm^{-1}), alkenes, C=CH_2, (3075–3095 cm^{-1}), and arenes (3010–3040 cm^{-1}), all of which present a more rigid framework against which the hydrogen atom vibrates, exhibit higher typical wavenumber values than the softer C–H stretches in alkyl and alkanal groups. Out-of-plane bending for arene and alkenic hydrogen atoms often gives characteristic absorption bands between 650–900 and 890–990 cm^{-1}, respectively.

Bands associated with the arene C–H stretch and out-of-plane bend appear clearly at about 3000 cm^{-1} and 680 cm^{-1}, respectively, in the infrared spectrum of benzene (Figure 2.37). Other prominent absorptions in this spectrum are due to the in-plane C–C bend (1040 cm^{-1}) and C–C stretching (1480 cm^{-1}) modes. For more complicated benzene derivatives, it is sometimes possible to determine the ring substitution pattern from differences in the characteristic absorptions in the infrared spectrum.

(ii) O–H, C–O bonds. Hydroxyl groups (OH) generally give a strong stretching band between 3590–3650 cm^{-1}. If the group is hydrogen-bonded there is usually a

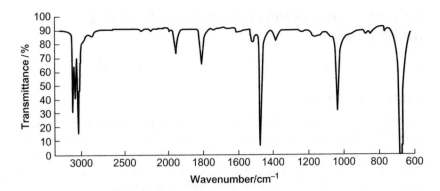

Figure 2.37 Infrared spectrum of benzene, C_6H_6 (liquid film).

broadening and a shift towards lower wavenumbers. For example, Figure 2.38 shows the infrared spectrum of ethanol, showing clearly the C–H and O–H stretches (2900 and 3300 cm^{-1}, respectively), absorption from aliphatic C–H bending (1400 cm^{-1}) and from C–O stretching (1050 cm^{-1}). The position and shape of the O–H stretching band is consistent with hydrogen-bonding in the liquid phase. Carbon–oxygen single bonds in esters, alkanols and ethers usually show a strong absorption in the range 1300–1050 cm^{-1}.

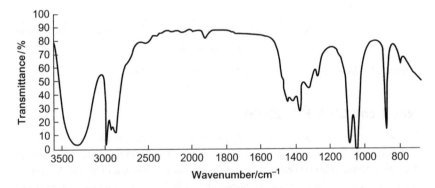

Figure 2.38 Infrared spectrum of ethanol, CH_3CH_2OH (liquid film).

(iii) N–H bonds. Like O–H bonds, N–H bonds show a characteristic absorption at high wavenumber (3500–3300 cm^{-1}) with any hydrogen-bonding again causing broadening and a shift towards lower wavenumbers.

(iv) C=O bonds. Carbonyl-containing compounds show a very characteristic strong C=O absorption in the range 1800–1600 cm^{-1}, the exact value of which depends on the structure of the adjacent groups. Simple ketones and aliphatic alkanals absorb at 1725 cm^{-1} (see Figure 2.36) but the wavenumber tends to be slightly higher for simple alkanoyl chlorides, esters, and anhydrides, and slightly lower than this for amides. The characteristic wavenumbers are also generally a little lower if the carbonyl group (in ketones, esters, etc.) is adjacent to an alkenic double bond (C=C–C=O) or to a phenyl ring, or if it is involved in hydrogen-bonding (see also Chapters 3 and 4).

2.15 Examples of infrared spectra of organic molecules

The following examples illustrate how infrared spectra vary with molecular structure.

Hexane, $CH_3(CH_2)_4CH_3$

The infrared spectrum of hexane (Figure 2.39) reveals very clearly the characteristic aliphatic C–H stretching, just below 3000 cm^{-1} (see also the 'Fuel' band labelled in the car exhaust gas infrared spectrum in Figure 2.30) and bending modes bands around 1500 cm^{-1}. No other functional group bands appear in the spectrum. The weaker bands to lower wavenumber are likely to be skeletal ring vibrations.

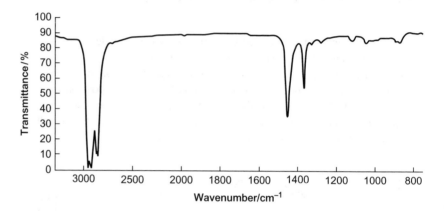

Figure 2.39 Infrared spectrum of hexane, $CH_3(CH_2)_4CH_3$ (liquid film).

Pent-1-ene, $CH_3(CH_2)_2CH{=}CH_2$

The pent-1-ene spectrum (Figure 2.40) shows some similarities with that of hexane with dominant features just below 3000 cm^{-1} and around 1400 cm^{-1} due to alkyl C–H stretching and bending modes. However, the higher wavenumber band at 3080 cm^{-1} due to the alkenic C–H stretching modes and at 900 cm^{-1} due to the out-of-plane

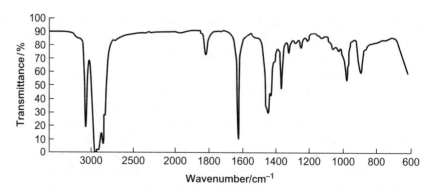

Figure 2.40 Infrared spectrum of pent-1-ene, $CH_3(CH_2)_2CH{=}CH_2$ (liquid film).

alkene C–H bending modes reveal the presence of the additional functional groups. The band at 1640 cm^{-1} is due to the C=C stretching vibration. Alkenes without a terminal double bond show slightly different absorptions below 1000 cm^{-1}, the nature of which can sometimes allow the substitution pattern of the alkene (*cis* or *trans*) to be established.

Methylbenzene, $C_6H_5CH_3$

The spectrum of methylbenzene (Figure 2.41) exhibits, in addition to bands associated with the aromatic and aliphatic C–H stretching modes (at 3050 and 2900 cm^{-1}, respectively), characteristic bands from the aliphatic C–H bending modes at *ca.* 1500 cm^{-1} and from out-of-plane aromatic C–H bending modes at *ca.* 700 cm^{-1}. The particular pattern around 700 cm^{-1} is typical of a mono-substituted benzene derivative.

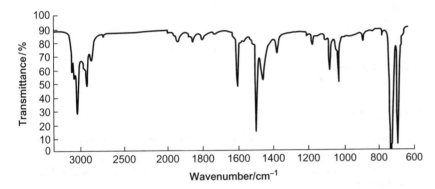

Figure 2.41 Infrared spectrum of methylbenzene, $C_6H_5CH_3$ (liquid film).

Ethanoic acid (acetic acid), CH_3CO_2H (thin liquid film)

The spectrum in Figure 2.42 from a thin film of the liquid, shows typical C–H bending vibrations (*ca.* 1400 cm^{-1}) and the characteristic carbonyl absorption at 1720 cm^{-1}. The broad absorption at 3000 cm^{-1} typifies a hydrogen-bonded OH group.

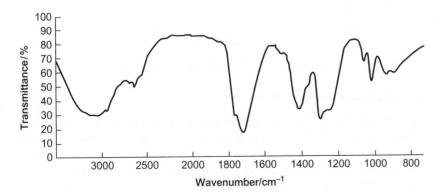

Figure 2.42 Infrared spectrum of ethanoic acid, CH_3CO_2H (liquid film).

Ethyl ethanoate (ethyl acetate), $CH_3CO_2C_2H_5$

The spectrum in Figure 2.43 shows clearly the typical carbonyl stretching absorption at 1740 cm^{-1}, appearing to slightly higher wavenumber than typically observed for ketones. The absorption at 1240 cm^{-1} is due to the C–O stretch, its prominence a particular characteristic for esters.

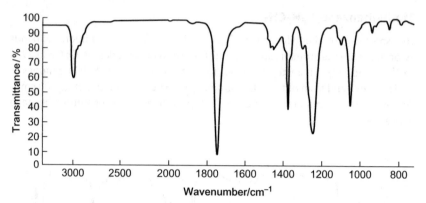

Figure 2.43 Infrared spectrum of ethyl ethanoate, $CH_3CO_2C_2H_5$ (liquid film).

Diethylamine, $(C_2H_5)_2NH$

The band at *ca.* 3300 cm^{-1} in the spectrum of diethylamine (Figure 2.44) is indicative of an N–H group with the broadening the result of inter-molecular hydrogen-bonding. The strong absorption at 1140 cm^{-1} is from a C–N stretching vibration (cf. C–O stretch in alcohols, ethers, etc.).

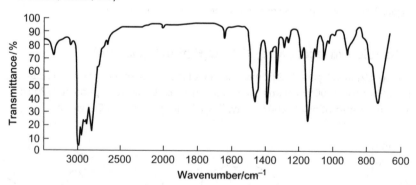

Figure 2.44 Infrared spectrum of diethylamine, $(C_2H_5)_2NH$ (liquid film).

Amides and the peptide link

Figure 2.45 shows the infrared spectrum obtained from a thin film of nylon-6,6 (a polymer prepared by the *copolymerization* reaction of hexane-dioic acid and hexane-l,6-diamine). Infrared absorptions from C–H stretching (2900 cm^{-1}), amide C=O stretching (1600 cm^{-1}) and N–H stretching (3300 cm^{-1}) help confirm the structure: note that the bonds around the amide function (the *peptide link* CO–NH) are planar; out-of-plane N–H bending accounts for the absorption at *ca.* 700 cm^{-1}.

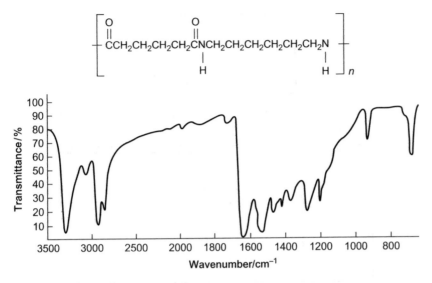

Figure 2.45 Infrared spectrum of nylon-6,6.

2.16 Carbonyl group modes in inorganic metal complexes

The application of infrared spectroscopy to the structural characterization of organic molecules relies to a large extent on general associations of bands in particular wave-number windows with particular functional groups. When used in the characterization of organic molecules, a great many of which are of low symmetry, the degree of finesse involved in deducing the structural features of a molecule relies on knowledge of how the positions of particular functional group bands depend on the local structural environment. In this respect, the deductions that one makes derive, to a large extent, from comparisons with, and experience of, the known spectra of related molecules. However, when dealing with molecules of high symmetry, which is often the case in inorganic chemistry where ligands arrange themselves in symmetrical arrangements around a metal centre, we can exploit the relationship between symmetry and the gross selection rules in infrared (and Raman) spectroscopy to deduce precisely how the ligands have arranged themselves when more than one possibility presents itself.

In the structural characterization of transition metal carbonyl complexes, the strongly absorbing CO stretch has been used to very good effect (see also Exercise 2.14). The strong intensity of the CO stretching bands and their distinctive placement away from other bands in the infrared spectra of such complexes both serve to facilitate their use in structural characterization. These features work exceptionally well in this particular context because the activity of the carbonyl stretching modes and the resulting number of distinct CO stretching bands will depend very much on how the ligands are arranged around the central metal atom. For example, the molybdenum carbonyl complex $Mo(CO)_4L_2$ (where L is a ligand which does not break the overall symmetry of the molecule) features four carbonyl ligands, arranged in one of two ways to produce either a *cis* or *trans* isomer. In each case, the four CO ligands will result in four possible CO stretching modes but their activity, the degree of any degeneracy and how the

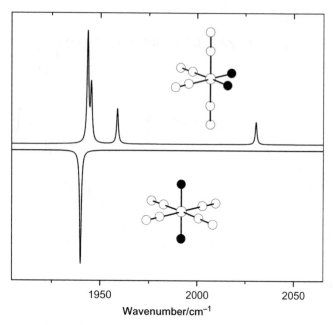

Figure 2.46 The infrared spectrum of *cis*-Mo(CO)$_4$L$_2$ (top) shows four carbonyl stretching bands where *trans*-Mo(CO)$_4$L$_2$ (bottom) shows just a single band.

resulting infrared spectrum will appear will depend on how those ligands are arranged around the central metal atom. The *cis* isomer has the lower symmetry, with all four CO stretching modes infrared active and all having different and distinct vibrational frequencies. Consequently, the CO stretch region of the spectrum exhibits four separately resolvable infrared bands (see Figure 2.46 (top)). The *trans* isomer on the other hand, having higher symmetry yields just a single infrared band (Figure 2.46 (bottom)). In this case, only two of the four CO stretching modes are IR active and these two modes are degenerate with one another, resulting in just a single band. The other two modes, which are not degenerate, are Raman active only (this molecule being subject to the mutual exclusion principle discussed earlier) and consequently do not appear in the infrared spectrum.

2.17 Summary

The broad scope of this chapter has presented a view of how spectroscopy can be used either as a high resolution tool to obtain direct measurements of key structural parameters in small molecules in the gas phase or as a tool to complement other techniques such as NMR and mass spectrometry in the structural characterization of larger molecules in the condensed phase. We have seen how the *rotational* energy level structure as revealed in pure rotational spectroscopy and in vibration/rotation spectroscopy yields rotational constants associated with different *moments of inertia* and how this information can be used to obtain *bond lengths* with high precision. In the condensed phase where rotational motion is restricted, both *infrared* and *Raman* spectroscopy yield *vibrational wavenumbers* associated with the different natural resonant frequencies of a molecule known as vibrational *normal modes*. In the case of larger molecules, some of those modes may be clearly associated with *functional groups* and the resulting bands that appear in an infrared or Raman spectrum can be used to deduce the presence of those groups within the molecule. *Symmetry* also plays a vital role in helping us to draw conclusions about the geometric arrangement of nuclei in a molecule. This is particularly useful in gas phase rotational and vibration/rotation spectroscopy but also in gas and condensed phase vibrational spectroscopy where molecules of higher symmetry display increasing selectivity in the activity of different modes in infrared spectra compared to Raman spectra. In such cases, the complementarity of the two techniques is particularly compelling.

We hope that by the end of this chapter you will appreciate and understand the following key elements:

- The relationship between rotational and vibrational energy level spacings and wavelength regions of the electromagnetic spectrum required to interrogate them

- The difference between optical absorption and emission spectroscopies and techniques relying on inelastic scattering to achieve changes in quantum state

- How to obtain a rotational constant from a rotational spectrum and consequently how to use that to find the related bond length

- An appreciation of the differences between r_e, r_0 and r_1 and the types of experiment we need to perform to obtain each parameter

- How rotational spectroscopy can be used to obtain information about the shape of larger molecules

- The basis of the quantization of vibrational energy in the context of the harmonic oscillator and the resulting selection rules

- The effect of anharmonicity on energy levels and selection rules in infrared and Raman spectroscopy

- The measurement of vibrational spectra—what distinguishes infrared from Raman spectroscopy?

- Vibrational degrees of freedom of a polyatomic molecule and normal modes of vibration

- The importance of symmetry in structural determination; the mutual exclusion principle in molecules with a centre of inversion

- Group vibrations and their importance in structural characterisation of organic and inorganic molecules.

2.18 Exercises

2.1. The rotational constant, B, obtained from the pure rotational spectrum of $H^{35}Cl$ is 10.4 cm^{-1}. Use Equations (2.2) and (2.5) to calculate the bond length, r, of $H^{35}Cl$.

2.2. Use Equation (2.8) to calculate the wavenumber positions of the first four lines in the rotational spectrum of $H^{35}Cl$. Convert these line positions into

Worked solutions to the exercises are available on the Online Resource

s^{-1}, expressing your answers in units of GHz or THz as appropriate, and then into wavelength. To which part of the electromagnetic spectrum do these wavelengths correspond?

2.3. In Worked example 2.1, a value of 1.132 Å was determined for the bond length of CO from the average adjacent line separations of the first 20 lines in the rotational spectrum. A similar calculation performed using only the first 8 lines of the CO rotational spectrum in Figure 2.6 yields a slightly shorter bond length of 1.131 Å. Suggest a reason why including a greater number of lines should yield a longer bond length.

2.4. Suggest an assignment for the very weak features appearing between about 20 and 60 cm^{-1} in Figure 2.6.

2.5. A far infrared spectrum of CO recorded at the coldest place on Earth is shown below.

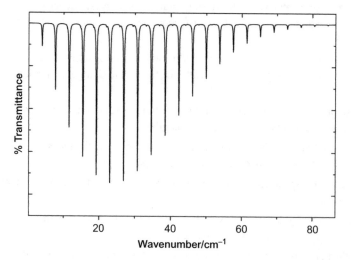

Use the spectrum together with Equation (2.15) to deduce the approximate temperature at the coldest place on Earth.

Note: The Boltzmann constant $k = 0.695$ cm^{-1} K^{-1} and the rotational constant, $B = 1.917$ cm^{-1}.

2.6. Given that the two equations (2.17) and (2.18) contain two terms each which differ only by a factor of m_D/m_H, solve the pair of equations (2.17) and (2.18) to show that $r_{12} = r_{CH} = 1.088$ Å and $r_{23} = r_{CO} = 1.110$ Å in HCO^+.

Note: $m_D = 2.0141$ u and $m_H = 1.0078$ u

2.7. The bond lengths of the linear triatomic molecule OCS were first determined in 1935 from electron diffraction (see Chapter 6) experiments as $r_{CO} = 1.16$ Å and $r_{CS} = 1.56$ Å. Use these values to predict the approximate position in MHz of the $J = 1 \rightarrow 2$ rotational transition in each of the two isotopologues $^{16}O^{12}C^{32}S$ and $^{16}O^{12}C^{34}S$.

Note: A useful conversion in calculating rotational constants from moments of inertia expressed in units of u.Å^2 is

$$B/MHz = \frac{505379.006}{I/u.\text{Å}^2}$$

The bond lengths obtained from those early electron diffraction experiments were used subsequently to assist in the search for the $J = 1 \rightarrow 2$ rotational line in the microwave spectrum of $^{16}O^{12}C^{34}S$. The line was successfully located in experiments reported in Physics Review in 1947 and the rotational spectra reported in that article were then used to obtain higher precision bond lengths for OCS. For additional background on the OCS story, see P.C. Cross and L.O. Brockway, *J. Chem. Phys.* **3**, 821 (1935) and T.W. Dakin, W.E. Good, and D.K. Coles, *Phys. Rev.* **71**, 640 (1947).

2.8. Calculate the number of vibrational modes in the following molecules:

   ammonia   ethyne   phenol   tryptophan C_{60}

2.9. Use Equation (2.21) to calculate the fundamental stretching frequency for a C–H bond, given that the atoms vibrate independently of other groups on the carbon atom and that the force constant is 4.8×10^2 N m^{-1}. You should assume that the effective mass, m_i, of the mode in this case is equivalent to the reduced mass, μ, of an independent C–H bond.

2.10. Use Equation (2.37) together with the values for B_e and α_e obtained in Worked example 2.7 for CO to compute the B_{10}, B_{20} and B_{30} rotational constants for CO and hence determine the percentage increase in the average bond length of CO in each of the $v = 10$, 20, and 30 vibrational levels compared to $v = 0$.

2.11. Account for the observation that in the infrared spectrum of deuterated benzene, C_6D_6, the C–D symmetric stretching mode occurs at *ca.* 2280 cm^{-1} where in undeuterated benzene the corresponding absorption is at 3050 cm^{-1}.

2.12. A thin film of perspex has strong absorptions at *ca.* 2950, 1730, 1450, and 1200 cm^{-1}. Use the information contained within Table 2.5 to suggest what type of polymer structure might be consistent with these observations.

2.13. The figure below shows the infrared spectrum of the compound whose NMR spectrum is the subject of Exercise 4.4, page 101 and whose mass spectrum is the subject of Exercise 5.3 (page 127). To what extent does the infrared spectrum support the assignments suggested by the mass spectrum and NMR spectrum?

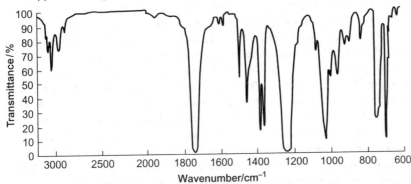

2.14. The infrared spectrum of the metal carbonyl compound $Fe_2(CO)_9$ exhibits carbonyl stretching bands at 2020 and 1830 cm^{-1}. By comparing the values of these two absorptions with the known wavenumbers of free carbon monoxide of 2146 cm^{-1} and of C=O stretches in organic carbonyls (see Table 2.5), suggest what insights the infrared spectrum might provide about the likely structure of this metal complex.

2.19 Further reading

L. M. Harwood and T. D. W. Claridge, (1997) *Introduction to Organic Spectroscopy*, Oxford Chemistry Primers, Oxford University Press, Oxford.

D. H. Williams and I. Fleming, (1995) *Spectroscopic Methods in Organic Chemistry*, 5th Edition, McGraw Hill, London.

J. Michael Hollas, (2004) *Modern Spectroscopy*, 4th Edition, Wiley, New York.

Electronic (ultraviolet–visible) absorption spectroscopy

3

3.1 Introduction

In addition to the absorption of well-defined amounts of energy to increase its vibrational and rotational energy, a molecule may also absorb energy to increase the energy of its **electrons**. The energy changes involved are considerably greater than those involved for vibration and rotation changes (described in the previous chapter) and correspond to radiation in the *ultraviolet* region (UV) (with wavelengths, λ, in the range of 200–400 nm) and *visible* region (λ 400–750 nm) of the electromagnetic spectrum.

It is also helpful to recognize that for radiation, of say, λ 400 nm, the frequency (ν) is 7.5×10^{14} Hz, and the corresponding energy of one quantum ($h\nu$) is 5×10^{-19} J. This is the energy which is absorbed by *one molecule* if it absorbs one quantum (or photon) of violet light. For one *mole* of material (in which the number of molecules is the Avogadro constant, N) the total energy absorbed (N quanta) corresponds to $(5 \times 10^{-19}) \times (6 \times 10^{23})$ i.e. *ca.* 300 kJ: this is the same order of magnitude as some *bond enthalpies* (energies) in typical molecules. This helps to explain the observation that interaction of visible or UV radiation with molecules sometimes brings about chemical reactions involving bond breakage: such processes are called **photochemical** reactions.

This chapter will introduce you to the principles which underlie the absorption of energy in the UV-visible region by electrons in molecules with double bonds (i.e. electronic energy changes) and the associated experimental arrangements. We will then provide examples of organic molecules which have characteristic absorption spectra in this region and associate this with key molecular features (e.g. double bonds, aromaticity, conjugation). We will explain how λ_{max} is determined, and related to structure, and show how the extent of absorption at a given wavelength (measured by the *molar decadic absorptivity*, or *extinction coefficient*, ε) depends upon the concentration of the sample (and also the molecular structure). Examples will be given which illustrate the effects of conjugation (e.g. with reference to the properties of indicators) and the consequent use of UV-visible spectroscopy in structural and quantitative analysis. We will conclude the chapter with examples to include inorganic systems and those of medical and biological relevance as well as some exercises.

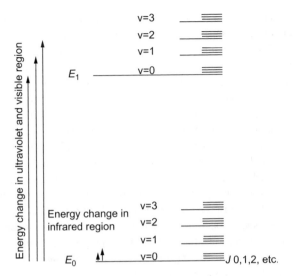

Figure 3.1 Representation of electronic, vibrational, and rotational energy levels.

3.2 Electronic energy changes

The overall energy of a molecule is the sum of contributions from electronic, vibrational and rotational energy, as given by Equation (3.1), where E_{vib} and E_{rot} have exactly the same allowed values as discussed previously (Chapter 2). This is also illustrated in Figure 3.1.

$$E = E_{elec} + E_{vib} + E_{rot} \qquad (3.1)$$

In a sample of a stable substance at room temperature, all the molecules will be in the E_0 level (the ground electronic state) but may have different v and J quantum numbers, as described in the previous chapter. When the appropriate energy is incident it is possible to excite the molecule to the higher energy *electronic* state E_1 (which again has similarly quantized rotational and vibrational energy levels). The excited state arises because of the promotion of electrons to higher energy *molecular orbitals*, exactly as can happen for electrons in atomic orbitals (*cf.* the sodium flame test, page 2, section 1.1).

This type of absorption explains why some organic and inorganic compounds are coloured—the absorption of energy for the transition $E_0 \rightarrow E_1$ is then in the visible region. Many other compounds absorb energy in the ultraviolet region and, although they do not appear coloured (unless there is also visible absorption), the absorption in this part of the spectrum can be detected. The related phenomena of fluorescence and phosphorescence, largely beyond the scope of this book, are associated with the re-emission of energy (particularly in the visible region) when the molecules in an excited electronic state return to the ground state.

3.3 Electronic absorption spectroscopy of organic molecules

The experiment

The approach is similar to that employed for infrared spectroscopy, and many commercial instruments are available, with different levels of cost and sophistication. The *source* provides radiation in a continuous range of wavelengths in the UV and visible region, as from a heated tungsten filament (e.g. an electric light bulb). Two separate sources are usually required for covering the whole UV and visible region. In a simple *dispersive* spectrometer a *prism* or *grating* is usually employed to separate (disperse) the radiation into component wavelengths, and the absorption of the sample at any particular wavelength is measured by the reduction of the signal from the detector (a *photoelectric device* (photomultiplier)) when the sample is placed in the incident beam.

Most investigations involve the use of liquid samples, which usually contain the compounds to be studied as solutions in suitable solvents. Many modern spectrometers employ a *double-beam* system whereby two identical beams of radiation are generated, one of which passes through the solution under investigation, the other of which passes through an equivalent amount of pure solvent, so that the *difference* in absorption which is measured is just that due to the molecules of the *solute* in the solution: the *solvent* should be transparent in the region of interest. Tetrachloromethane, hexane, cyclohexane, and ethanol are often employed as solvents, and the solutions are usually contained in quartz glass cells such that the beam of radiation passes through 1 cm of solution (this is known as the *path length*).

Most spectrometers automatically record *absorption* as a function of *wavelength*. In very simple spectrometers the operator may record the absorption with separate measurements at different wavelengths. (A colorimeter is a simple type of spectrometer working with a given colour; i.e. at a fixed range of wavelength in the visible region, selected with a suitable filter from a wide range of radiation.)

In a typical *photodiode-array* spectrometer a rather different arrangement is employed. The sample (prepared as described above) is irradiated with white light (i.e. irradiated simultaneously with all the appropriate wavelengths in the UV-visible range) before the resulting beam is then passed through a prism (or onto a grating) to separate out the individual wavelengths (some of which may have been partially absorbed): the resultant beams then fall simultaneously on an array of photodiodes, each of which measures the intensity of the beam at an individual wavelength. This offers a relatively rapid way in which a complete spectrum (extent of absorption vs. wavelength) can be recorded quickly; it can be particularly effective when coupled to a multi-channel approach to obtain spectra rapidly from a host of samples (e.g. in biological or medical testing applications).

Examples of spectra

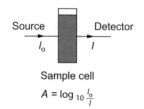

$$A = \log_{10} \frac{I_0}{I}$$

Figures 3.2–3.4 are the electronic absorption spectra of three organic molecules—propanone, benzene, and the indicator methyl red, for each of which the concentrations are given. These are plots of *wavelength* (λ/nm) against *absorbance* (A) (sometimes also referred to as *optical density*). The absorbance is the logarithm of the ratio of the intensity of the incident radiation (I_0) to that of the transmitted radiation (I) (note that reference will sometimes be made to transmittance, T, which is defined as I_0/I).

A peak in the spectrum at a given λ corresponds to absorption of energy at this wavelength by the solute molecules; for some molecules, more than one area of absorption is observed, as in Figures 3.2 and 3.3: strong absorptions tailing off into the normally inaccessible region below 200 nm are detected in addition to the higher-wavelength peaks. Propanone and benzene are colourless, since they do not absorb in the visible region, but methyl red absorbs in acid solution in the blue-green region

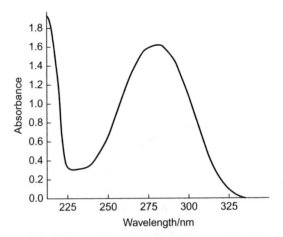

Figure 3.2 Electronic absorption spectrum of propanone, $(CH_3)_2C=O$ (for a solution of propanone in hexane, of concentration 6 g dm^{-3}, using a 1 cm cell).

(λ 400–600 nm) and so appears red; in alkaline solution, the absorption is at lower wavelengths and the solution is yellow. The peaks observed are broad in most cases because molecular interactions in the liquid cause the obliteration of the expected vibrational and rotational fine structure, although separate peaks (associated with vibrational fine-structure) are clearly apparent in the absorption with λ *ca.* 250 nm from benzene in Figure 3.3.

Electronic absorption spectra are usually characterized by two parameters:

(i) the values of the *wavelengths* at which *absorption maxima* occur (λ_{max});
     Figures 3.2–3.4 indicate the different values obtained for the different molecules

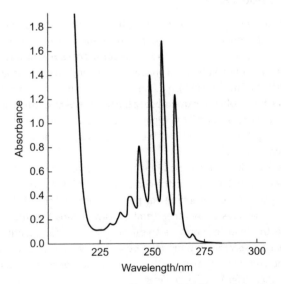

Figure 3.3 Electronic absorption spectrum of benzene, C_6H_6 (for a solution in hexane, of concentration 0.6 g dm^{-3}, using a 1 cm cell).

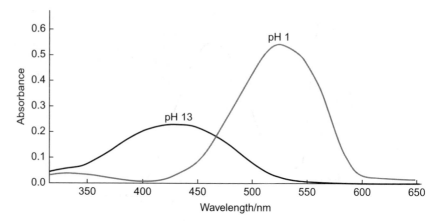

Figure 3.4 Electronic absorption spectra of aqueous solutions of the indicator methyl red (containing 0.004 g dm^{-3} of the indicator; recorded with a 1 cm cell).

considered; for example, for propanone, the high-wavelength absorption has $\lambda_{max} = 279$ nm; and

(ii) the *extent* of absorption, for a given concentration of compound, at any given wavelength (i.e. the *height* of the peak).

As will be seen, the position (λ_{max}) and extent of absorption provide two more characteristic properties of a molecule which depend on its structure and hence, as with infrared and Raman spectroscopy (Chapter 2) and NMR spectroscopy (Chapter 4), are helpful in diagnosis and problem-solving.

The Beer–Lambert Law

The extent of absorption at a given wavelength by an absorbing compound in a non-absorbing solvent is found to depend upon the *concentration* of the compound (c) and upon the *path length* of the cell (d). The **Beer–Lambert Law**, which is generally well obeyed for fairly dilute solutions, expresses the dependence of the absorbance on these two variables (Equation (3.2)). If d is in m and c is in mol m^{-3}, then ε, in m^2 mol^{-1}, is described as the **molar decadic absorptivity** or **molar extinction coefficient**. Thus if d and c are known, and if the experiment gives a value for A (the spectrometer records this), ε can be calculated. This is usually quoted for the wavelength of maximum absorption (i.e. at λ_{max}).

Once ε has been determined, the value of A for a given solution of a known compound can be used to determine c, the concentration of that compound in the solution (assuming that no other ingredient present absorbs light of this wavelength). This behaviour, the basis of the Beer–Lambert Law, means that measurement with UV-visible spectrometers can provide an excellent method for quantitative as well as qualitative analysis (*spectrometry* and *spectrophotometry* are expressions that are often employed when reference is made to the determination of concentrations of known components, often referred to as *analytes*).

Let's now move on to a pair of worked examples which illustrate simple calculations relating ε and c.

$$A = \log_{10} I_0 / I = \varepsilon c d \qquad (3.2)$$

where A is the absorbance, as measured by the spectrometer;
d is the path length of the cell;
c is the concentration of the absorbing species (analyte)
ε is a constant for a particular compound at chosen wavelength

Worked example 3.1

Question. Figure 3.2 is the UV spectrum of a solution of propanone in hexane of concentration 6 g dm^{-3}, using a 1 cm cell. Calculate the molar decadic absorptivity (in m^2 mol^{-1}) for the peak with $\lambda = 279$ nm.

Answer. First, examine Figure 3.2 and identify the peak at λ 279 nm. Note that $A = 1.60$ for this solution. Then insert this value into the Beer–Lambert Law ($A = \varepsilon cd$) together with d in metres (i.e. 10^{-2}, since it is a 1cm cell) and c in moles dm^{-3} (convert the given concentration to this molar concentration from the given concentration of 6 g dm^{-3}, i.e. 6/58 mol dm^{-3}, that is 0.103 mol dm^{-3}, i.e. 0.103×10^3 mol m^{-3}). Thus $\varepsilon = A/cd = 1.60/0.103 \times 10^3 \times 10^{-2} = 1.55$ m^2 mol^{-1}.

Worked example 3.2

Question. For a solution of acetophenone ($C_6H_5COCH_3$) in hexane, for which the UV-visible spectrum is shown in Figure 3.6, the molar decadic absorptivity of the peak with λ_{max} 279 nm is 100 m^2 mol^{-1}. Calculate the concentration of the solution. Compare your answer with the concentration of benzene used in Figure 3.3.

Answer. First, examine the UV–visible spectrum of acetophenone (Figure 3.6, page 61) and note the peak at λ 279, for which it is given that ε is 100 m^2 mol^{-1}. Note that the absorbance A is 0.1. Then use the Beer–Lambert Law $A = \varepsilon cd$, so that $0.1 = 100 \times c \times d$, with $d = 10^{-2}$ m, (i.e. a 1 cm cell) giving c (in mol m^{-3}) as $A/\varepsilon d$, i.e. $0.1/100 \times 10^{-2}$, i.e. 0.1 mol m^{-3}. Next take this concentration (0.1 mol m^{-3}) and convert to moles dm^{-3} (10^{-4} mol m^{-3}) and then into g dm^{-3} (with the molecular weight of acetophenone as 120) as 120×10^{-4} i.e. 0.012 g dm^{-3}.

Note that this concentration is much less than that employed to get an absorption of approximately similar intensity (absorbance) from benzene (Figure 3.3), reflecting the much higher value of ε for the more conjugated molecule (20 for benzene; >1000 for acetophenone); the reason for this is explained in the next section.

3.4 The relationship of λ_{max} and ε_{max} to structure

The electronic absorption spectra of a variety of organic compounds show that only certain types of molecule exhibit absorption in the UV-visible range (λ 200–750 nm). These are found to contain double or triple bonds (and, in some cases, lone-pairs of electrons), which are essentially responsible for the absorption; these fragments are called **chromophores**. When two or more chromophores are adjacent to each other (the groups are then said to be **conjugated**) the absorptions become more pronounced (with higher ε_{max}) and occur at lower energy (with greater λ_{max}).

This behaviour can be understood in terms of the types of molecular orbitals involved in the electronic excitations. There are three different types of molecular orbital, and the electrons in these orbitals have somewhat different environments. First, there are the electrons in the σ-orbitals which constitute the bonding framework of a molecule (e.g. the electrons in the 4 C–H bonds in methane, CH_4). These orbitals are formed from overlap of s, sp, sp^2, and sp^3 orbitals. Second, there are electrons in π-orbitals, formed from laterally overlapping atomic p-orbitals in compounds such as benzene and ethene (ethylene). Third, there are *lone-pair* electrons in orbitals on atoms like oxygen, nitrogen, etc.; these are called non-bonding

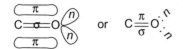

or n-electrons. The carbonyl group (in a ketone, say) contains all three types of molecular orbital.

When two electrons in atomic orbitals are brought together to form a bond (i.e. when a filled molecular orbital is produced) a higher energy *anti-bonding* orbital is also formed which is empty in the ground state of the molecule (*cf.* empty high-energy atomic orbitals). When excitation takes place, an electron from one of the filled orbitals (σ, π, or n) becomes excited to a vacant anti-bonding orbital (σ^*, π^*) so that a new *excited* state is reached (an example, involving the excitation of π-electrons in ethylene, is shown below). Since various excitations are possible, depending on the structure of the molecule, there may be various possible absorptions, corresponding to the transitions $n \rightarrow \sigma^*$, $\sigma \rightarrow \pi^*$, etc.

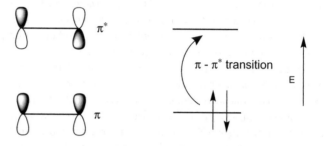

π and π^* molecular orbitals for ethene, showing the $\pi \rightarrow \pi^*$ transition.

The *approximate* relative energies of typical σ, π, n, and anti-bonding orbitals are indicated in Figure 3.5; as expected, the σ-electrons are the most-tightly bound (most energy is needed to excite them). The order of decreasing energy for the absorptions is as follows:

$$\sigma \rightarrow \sigma^* > \sigma \rightarrow \pi^* \sim \pi \rightarrow \sigma^* > \pi \rightarrow \pi^* \sim n \rightarrow \sigma^* > n \rightarrow \pi^*$$

Of all these possible transitions, only those of the last three types normally account for absorption in the UV-visible region, the others requiring too great an energy. This then explains why only molecules with n or π electrons give rise to characteristic UV and visible spectra, whereas alkanes, for example, show no absorption in this region. Note that ethanol absorbs radiation of wavelength *ca.* 200 nm and below. This absorption derives from the $n \rightarrow \sigma^*$ absorption by the lone-pair electrons on the oxygen atom. However, ethanol is transparent above this wavelength and finds use as a solvent for UV-visible studies on molecules with higher-wavelength absorptions.

Let's now consider a variety of other organic molecules which contain chromophores (double bonds, etc.) and which give absorption in the UV-visible region; Table 3.1 summarizes the measured values of ε_{max} and λ_{max}, together with the types of transition involved.

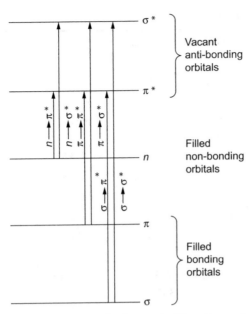

Figure 3.5 Approximate relative energies for electrons in different types of molecular orbital in organic compounds (not to scale).

For example, for propanone (see Figure 3.2) the peak at λ = 188 nm (part of which is shown and for which ε = 90 m^2 mol^{-1}) is responsible for absorption at the low-wavelength end of the observed spectrum, and the peak at λ = 279 nm (ε = 1.5 m^2 mol^{-1}) is also clearly visible. These are due to transitions involving the electrons in the *carbonyl-group double bond* ($\pi \to \pi^*$) and the *oxygen's lone-pair electrons* ($n \to \pi^*$), respectively. The characteristic values of λ_{max} and ε_{max} are slightly sensitive to the solvent used.

These absorptions do not depend to any marked extent on the nature of the alkyl groups attached to the carbonyl function; thus for a variety of ketones and alkanals, the UV spectra recorded for hexane solutions show the following values of λ and ε for the $n \to \pi^*$ absorption: butanone (279 nm, 1.6 m^2 mol^{-1}), cyclohexanone (285 nm, 1.4 m^2 mol^{-1}), ethanal (acetaldehyde, 293 nm, 1.2 m^2 mol^{-1}), propanal (290 nm, 1.8 m^2 mol^{-1}). For carbonyl-containing compounds of different chemical type (e.g. alkanoate esters), the absorptions are, however, characteristically different (see Table 3.1).

When a *carbon–carbon double bond* is present, a characteristic $\pi \to \pi^*$ absorption is observed (Table 3.1 gives data for pent-1-ene: see also the diagram on page 58). What is particularly noticeable here is the larger value of ε_{max}, which means that a more dilute solution of the compound is needed, compared with the ketones, to obtain the same amount of absorption. This type of increase often occurs when a transition takes place between states of similar type (e.g. $\pi \to \pi^*$ compared with $n \to \pi^*$; the former transition is said to be **allowed**).

When two chromophores in a molecule are adjacent (conjugated) it is generally found that the energy needed for absorption *decreases* (i.e. λ_{max} increases) and the

Table 3.1 Characteristic UV-visible absorptions for organic compounds.

(a) **Molecules with single chromophores**		Solvent	λ_{max}/nm	Transition type	ε_{max}/m^2 mol^{-1}
Propanone	$(CH_3)_2CO$	hexane	188	$\pi \rightarrow \pi^*$	90.0
			279	$n \rightarrow \pi^*$	1.5
Ethyl ethanoate (ethyl acetate)	$CH_3C(O)OCH_2CH_3$	water	204	$n \rightarrow \pi^*$	6.0
Pent-1-ene	$CH_3CH_2CH_2CH=CH_2$	hexane	190	$\pi \rightarrow \pi^*$	1000
Nitromethane	CH_3NO_2	hexane	278	$n \rightarrow \pi^*$	1.7
(b) **Conjugated molecules**					
Buta-1,3-diene	$CH_2=CH-CH=CH_2$	hexane	217	$\pi \rightarrow \pi^*$	2100
Butenone	$CH_2=CH-C(O)CH_3$	ethanol	219	$\pi \rightarrow \pi^*$	360
			324	$n \rightarrow \pi^*$	2.4
Benzene	C_6H_6	hexane	184	all $\pi \rightarrow \pi^*$	6000
			203		740
			255		20
Acetophenone	$C_6H_5C(O)CH_3$	ethanol	199	all $\pi \rightarrow \pi^*$	2000
			246		1260
			279		100
			320		4.5
Nitrobenzene	$C_6H_5NO_2$	hexane	252	all $\pi \rightarrow \pi^*$	1000
			280		100
			330		12.5

extent of absorption (ε_{max}) *increases* compared with the values for separate groupings. This is illustrated for a molecule with two *conjugated* double bonds (e.g.1,3-butadiene): see Table 3.1 and the transition indicated on the diagram on the next page, which shows the reduction in separation between the appropriate π and π^* orbitals. This effect is also notable for the conjugation between adjacent C=C and C=O groups (see Table 3.1) in butenone (*cf.* propanone). The effect is particularly marked for benzene (and other aromatic compounds) which have extended and cyclic π-systems. These molecules can also be distinguished since there are several possible $\pi \rightarrow \pi^*$ absorptions (owing to the existence of several π and π^* orbitals).

When a benzene ring and another chromophore are conjugated then the characteristic absorptions are shifted to even longer wavelengths: an example is provided by Figure 3.6, which shows the electronic absorption spectrum of acetophenone, $C_6H_5COCH_3$, which was referred to earlier in Worked example 3.2 (there is also an even weaker absorption at higher wavelength; the details are given in Table 3.1). Also note that substitution of alkyl groups on aromatic and alkenic chromophores leads to small increases in λ_{max}.

The increased wavelength of absorption for *conjugated* molecules (i.e. when chromophores are adjacent to each other) can lead, if enough chromophores are

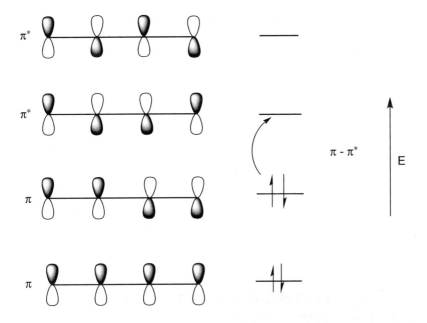

π and π^* orbitals for buta-1,3-diene, showing the lowest energy $\pi \rightarrow \pi^*$ transition

present, to an absorption in the visible region. This occurs, for example, for aromatic compounds containing fused rings, for some 1,2-diketones (which are yellow), for large molecules (with delocalized π-electrons) used as pH indicators (the conjugation and hence the colour depends upon the ionization of groups in the molecule), and for molecules containing chains of double bonds.

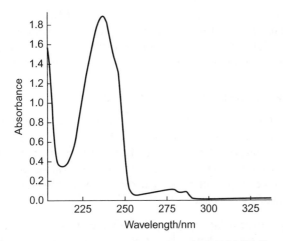

Figure 3.6 Electronic absorption spectrum of acetophenone, $C_6H_5COCH_3$ (for a solution in hexane using a 1 cm cell).

For example, β-carotene (which occurs in carrots) is orange and has $\lambda_{max} = 450$ nm, $\varepsilon_{max} = 15{,}000$ m^2 mol^{-1}. When ε_{max} is very high, as in this case, only very dilute solutions are needed for detection of the absorbing molecules, and the technique becomes a very sensitive method indeed for the detection of such absorbing species (for example, absorption could be detected for a solution made up from less than 0.01 mg of β-carotene).

3.5 Some applications of UV and visible absorption spectroscopy

Structural analysis

One important use of electronic absorption spectroscopy is the recognition of chromophores or groups of chromophores in organic molecules, by the measurement of λ_{max} and ε_{max} for the various absorption peaks. This information usually allows the *type* of molecule to be determined and, particularly when used in conjunction with other spectra, may provide valuable assistance with the determination of the exact molecular structure.

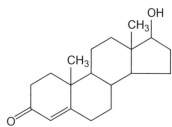

This branch of spectroscopy has been particularly useful in the structural investigation of steroids, which are biologically important molecules. An example is the hormone testosterone, whose structure is shown in the margin. The absorption at 241 nm ($\varepsilon = 1600$ m^2 mol^{-1}) is characteristic of adjacent C=C and C=O double bonds in this type of cyclic structure; it has proved possible in examples like this to use the UV data to give an indication of the detailed structure around the chromophores, though not for more distant parts of the molecule (the ^{13}C NMR spectrum of the structurally related molecule cholesterol acetate, which provides much more information, is given on page 93, section 4.7)

Quantitative analysis (spectrophotometry)

We can often use the Beer–Lambert Law to relate the absorbance from a particular substance to its concentration (if ε is known), with many applications in quantitative analysis. The advantages of this method are that very low concentrations can be reliably obtained and that the *rate* of change of the absorption at a given wavelength can easily be monitored (e.g. if the compound is involved in a chemical reaction). Some typical applications are as follows.

(i) *Kinetic Investigations.* Most spectrometers produce a plot of absorbance against λ but also allow the absorbance at one particular wavelength to be plotted as a function of time. This facility leads to a plot of concentration against time for either the loss of a reactant or the build-up of a product, as long as the absorption is characteristic of only the single component under investigation. Thus, the spectrometer not only provides vital information about the nature of the product (or products) of a reaction, it may also allow the rate of the reaction to be followed. Both pieces of information can aid the elucidation of the reaction mechanism concerned.

(ii) **Keto–enol tautomerism.** Both the UV spectrum (Figure 3.7) and the ^{1}H NMR spectrum of pentane-2,4-dione (acetylacetone, $CH_3COCH_2COCH_3$), to be discussed later (page 86, section 4.5), indicate that this molecule does not exist simply as the molecular formula suggests—i.e. with two (independent, unconjugated) keto groups. The NMR spectrum provides detailed evidence for two different forms of the molecule present in equilibrium: these are the **keto** and **enol tautomers** of the molecule (the phenomenon is called **tautomerism**).

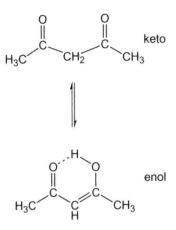

Figure 3.7 shows the absorption for 10^{-4} mol dm^{-3} solutions of acetyl acetone in hexane, ethanol, and water. First, the extent of absorption suggests that a simple carbonyl-type structure is not responsible for the peak (for example, a much greater concentration of propanone is necessary to obtain a strong ($n \to \pi^*$) absorption: see Figure 3.2): a $\pi \to \pi^*$ transition associated with a *conjugated* structure is more likely. Second, the variation in the extent of absorption with solvent is surprising (it doesn't occur for a simple ketone).

The explanation is that the absorption at $\lambda = 270$ nm is due to the *enol* form of the molecule (*cf.* absorptions for compounds of the type C=C–C=O, Table 3.1) and it can be demonstrated (e.g. by ^{1}H NMR spectroscopy: see page 86, section 4.5) that for a solution in ethanol there is approximately 73% of the enol and 27% of the keto form together in equilibrium (the percentage of the former accounts for the absorption with $\lambda_{max} = 270$ nm, absorbance = 0.96 from the 10^{-4} mol dm^{-3} solution of the diketone in ethanol). The absorptions for the other solvents imply that in water there must be a much smaller proportion of enol than when ethanol is the solvent and that in hexane solution there is correspondingly more enol. In hexane, the internally hydrogen-bonded enol form is preferred (the dotted line in the structure indicates a

Figure 3.7 Electronic absorption spectra of solutions (10^{-4} mol dm^{-3}) of pentane-2,4-dione, $CH_3COCH_2COCH_3$, in hexane (—), ethanol (–) and water (---); the cell has a path length of 1 cm (see Exercise 3.2).

hydrogen bond) whereas in aqueous and alcoholic solutions the formation of *inter-molecular* hydrogen bonds between the carbonyl group and the water molecules (or alcohol molecules) stabilizes the keto form.

Indicators

Figure 3.8 shows the variation in the absorption spectrum of a dilute aqueous solution of the indicator methyl red as the pH is altered. The indicator is red in acid solution (λ_{max} = 520 nm, see Figures 3.4 and 3.8) and yellow in alkali (λ_{max} = 425 nm), these being the colours of the acid and base forms which differ in the extent of conjugation.

At pH 1, the indicator is essentially all in the acid (HA) form; at pH 13, it is essentially all in the base (A⁻) form. For intermediate pH values, both HA and A⁻ are present (to an extent which is governed by the pH and by the dissociation constant of the indicator, K_a). The heights of the 'acid' peak (λ = 520 nm) and the 'base' peak (λ = 425 nm) can be used, together with the 100% 'acid' and 'base' absorptions, to determine the concentrations of HA and A⁻ at any given pH. This then leads to a measurement of K_a. Alternatively, if K_a is known, then measurement of [HA] and [A⁻] gives a value for the pH. As will be appreciated, these measurements serve to quantify the procedure whereby, in a titration, the eye responds to the change in colour of a solution when a predominance of HA, say, is changed to a predominance of A⁻.

Note in this case the *isosbestic point* through which all the absorption spectra pass. Its occurrence is characteristic of systems with two different chromophores which are interconvertible and in equilibrium.

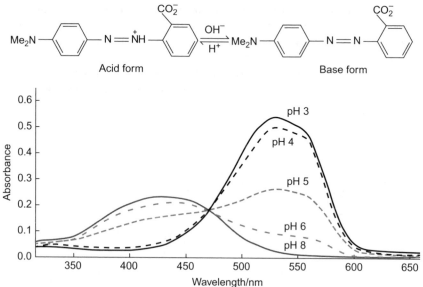

Figure 3.8 Electronic absorption spectra of aqueous solutions of methyl red (all containing 0.004 g dm⁻³ of the indicator) at various pH values.

Metal ions and complexes

The application of UV-visible absorption spectroscopy also allows the determination of absorption maxima and molar decadic absorptivities for inorganic ions with electronic absorptions in this part of the spectrum (e.g. for the absorptions which account for the purple colour of the manganate (VII) ion (permanganate, MnO_4^-) and for the yellow colour of the dichromate (VI) ion ($Cr_2O_7^{2-}$). This type of detailed information can then be useful in several contexts. For example, the ions MnO_4^- or $Cr_2O_7^{2-}$ could be identified as present in a solution from the appearance of their characteristic absorption spectra and their concentrations measured if there are no other competing absorptions present. Further, the inorganic anions mentioned above could be formed by the oxidation of solutions containing trace quantities of, say, Mn(II) and Cr(III) whose concentrations themselves are to be determined: the extent of absorption (the absorbance) of the oxyanions at the appropriate wavelengths can be used to calculate the concentrations of the ions present (this can often be done for individual ions and for mixtures of ions from a single spectrum).

These approaches underpin, of course, the traditional use of different colour tests, or 'spot tests' to recognize the presence of different elements in an 'unknown' sample (and, by extension, to determine their concentration using the Beer–Lambert Law where appropriate). A good example is the use of the complexing agent dimethyl glyoxime to reveal the presence of Ni^{2+} by the production of the characteristic red complex (e.g. of an ethanolic solution); in a gravimetric determination of the amount of metal ion present, a precipitate would be removed by filtration and weighed.

There are many ingenious examples of the design of special sensors for individual metal ions which can be both extremely sensitive and also selective for one specific ion rather than others (e.g. in the same period or group). In some cases this is done by including a relatively specific (but not chromophoric) group with potential for bonding, attached to the absorbing molecule which then 'reports' on the capture of a specific ion (in the example shown, capture of sodium ions, for example, by the oxygen and nitrogen atoms in the cyclic ether causes a colour change from orange to yellow in the attached chromophore.)

R = Re(CO)$_3$(2,2-bipyridine)

In addition, there is considerable interest in the detailed analysis of absorptions from metal ions and complexes in terms of the electronic structure (and hence the possible electronic transitions) of the molecules or ions concerned. Particularly useful information can be obtained about transitions and energy levels involving d-electrons in complexes containing transition-metal ions.

Biological and medical applications.

Details of the very wide range of applications of the principles of UV-visible spectrophotometry in biological applications are outside the scope of this book but we can note that these include, for example: recognition of DNA and RNA from their characteristic absorptions around 260 nm (associated with the cyclic, conjugated purine and pyrimidine bases), and proteins from the characteristic aromatic amino-acid absorptions at *ca.* 280 nm; use of specific reagents which produce characteristic colour change as a result of complexation (e.g. the Biuret test for proteins); determination of enzyme kinetics by following the build up or decay of specific products of enzyme reaction with characteristic λ_{max}. And medical applications can be illustrated with reference to the determination of the concentrations (and recognition) of molecules with characteristic absorptions including porphyrins (e.g. in urine) and haemoglobin, and applications of immunoassays in which, for example, a tumour marker (or other macromolecule referred to as an *antigen*, often a particular protein) is recognized via its highly specific bonding to an *antibody* which itself has been designed with a chromophore attached (and hence which 'reports on' the binding, allowing detection and quantification of the antigen). Other related clinical immunoassays based on colorimetric tests (and enzyme linkage) can be used to determine insulin (to assess hypoglycaemia), and prostate-specific antigen (PSA) used in the diagnosis of prostate cancer.

3.6 **Summary**

As a result of studying the material presented in this chapter you should be able to appreciate the principles which underpin the theory and application of electronic absorption spectroscopy and its use in structural and quantitative analysis.
You should now be able to:

- predict the types of molecule for which electronic transitions (e.g. $n \rightarrow \pi^*$, $\pi \rightarrow \pi^*$) give rise to absorption in the UV-visible region;

- understand, in broad terms, the relationship between λ_{max} (and ε_{max}) and structure;

- employ the Beer–Lambert law to calculate concentrations of absorbing species in solution from measurement of absorbance;

- use UV-visible results, especially in conjunction with infrared and NMR spectroscopy (and mass spectrometry) to solve structural problems.

You should also be able to appreciate the potential use of UV–visible spectroscopy in biological and medical analysis.

3.7 **Exercises**

Worked solutions to the exercises are available on the Online Resource

3.1 Figure 3.3 was recorded for a solution of benzene in hexane of concentration 0.6 g dm^{-3}, using a 1 cm cell. Calculate the molar decadic absorptivity (m^2 mol^{-1}) for the peak with $\lambda = 255$ nm.

3.2 (a) Calculate the *apparent* molar decadic absorptivity for λ_{max} for pentane-2,4-dione in ethanol from Figure 3.7, which shows the spectrum for a solution containing 10^{-2} g dm^{-3} of the di-ketone.

(b) The proportion of the enol of pentane-2,4-dione in solution in ethanol has been estimated as 73%. From Figure 3.7 calculate the proportion of enol for solutions in (i) hexane, and (ii) water, and comment on the result.

3.3 Explain why aniline (phenylamine, $C_6H_5NH_2$) shows absorption maxima (λ_{max}) at approximately 230 and 280 nm ($\varepsilon = 860$ and 143 $m^2 \, mol^{-1}$, respectively), whereas salts of the anilinium (phenyl-ammonium) cation ($C_6H_5NH_3^+$) have absorptions at *ca.* 200 and 250 nm ($\varepsilon = 750$ and 16 $m^2 mol^{-1}$, respectively).

3.4 From Figure 3.8, estimate the dissociation constant K_a of methyl red.

3.8 Further reading

L. M. Harwood and T. D. W. Claridge, (1996) *Introduction to Organic Spectroscopy*, Oxford University Press, Oxford.

D. H. Williams and I. Fleming, (2007) *Spectroscopic Methods*, 6th Edition, McGraw-Hill, London.

4 Nuclear magnetic resonance spectroscopy

4.1 Introduction

Since the first nuclear magnetic resonance (NMR) experiment was successfully demonstrated in 1946, this spectroscopic technique has become so powerful that it plays an essential role in modern research. Indeed, NMR and mass spectrometry (described in the next chapter) have no real rivals as the approaches most widely applicable for the solution of structural problems, and their use has now become routine. In the form of MRI (magnetic resonance imaging), NMR has also moved from the realm of research to become an essential diagnostic method in hospitals and a feature of modern medicine.

This chapter will firstly introduce you to the concept of nuclear spin, and hence nuclear magnetic moments and their study by NMR spectroscopy. Using the proton, ^{1}H, as the simplest example, we will show how the phenomenon arises, and describe the operation of a basic NMR spectrometer. With emphasis on examples chosen from a range of organic molecules, we will show how electronic effects of different substituent groups enable these to be recognized by the chemical shifts of the protons concerned, leading to the use of the technique as a diagnostic tool. The occurrence of spin–spin splittings will be interpreted in order to show how neighbouring protons interact with each other, to provide further structural information; opportunities will be provided for practice with worked examples and further problems. Examples of tautomerism, hydrogen-bonding, and dynamic effects within molecules will also be described.

We will then discuss the development of Fourier Transform instrumentation and its application to NMR spectroscopy, illustrated with examples of other magnetic nuclei, and special emphasis will be placed on the structural information about organic molecules especially that which can be derived from ^{13}C NMR spectroscopy (and with further problems for practice). In the final section, we will discuss the development of 2-dimensional NMR spectra and then outline the principles and uses of MRI in medically-related studies.

4.2 The NMR experiment

At the heart of each NMR experiment lies the nucleus of an atom. An atom can be thought of as a sea of negatively charged electrons surrounding a positively charged nucleus which is itself composed of protons (positively charged) and neutrons. It has been found that nuclei which contain an *odd* number of protons or an *odd* number of neutrons (or odd numbers of both) possess an extra property, in addition to their charge, which can be demonstrated by a suitable experiment. They can be shown to have a **magnetic moment** which means that in the simplest case they behave like tiny bar magnets. This phenomenon can be demonstrated by detecting the energy of interaction when the nuclei are placed between the pole-pieces of a magnet. As happens when a simple bar magnet is placed in the magnetic field of a second magnet, the distinction between attractive and repulsive interaction can be detected.

The hydrogen atom, ^1H, has a single proton for its nucleus and hence has a magnetic moment. In the presence of an externally applied magnetic field, the magnetic moment experiences an interaction which results in it becoming aligned either *parallel* to or *opposed* to the direction of the applied field. This picture of two allowed orientations is also valid for ^{13}C, ^{19}F, ^{31}P, and ^{15}N, but for nuclei like ^2H the situation is more complex.

The two possible alignments of the magnetic moment of the hydrogen nucleus, ^1H, are represented diagrammatically in Figure 4.1. These arrangements have different *energies*, and an exact amount of energy is necessary to twist the magnet from one position to the other, i.e. from the attractive to the repulsive situation.

> It will be helpful to remember that a magnetic moment (or field) is associated with a body which is charged and in motion (cf. the magnetic field from electrons moving in a wire). This requires the ^1H nucleus to be spinning in one of two directions.

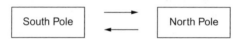

Figure 4.1 The two allowed alignments of the nuclear magnetic moment of a hydrogen atom in an applied magnetic field.

Now, for a magnetic moment, μ, the component in the direction of the applied field is $+\mu$ (designated α) or $-\mu$ (designated β), depending on whether it is parallel or antiparallel to it. The energies of these two arrangements of the magnet are $-\mu B_0$ (aligned) and $+\mu B_0$ (opposed) respectively, where B_0 is the magnetic flux density of the applied field. The energy difference between the two arrangements is $2\mu B_0$ and this amount of energy (as radiation of the corresponding frequency) is necessary to invert the proton's magnetic moment from the position of lower energy to that of higher energy.

The exact condition is then:

$$\Delta E = h\nu = 2\mu B_0 \tag{4.1}$$

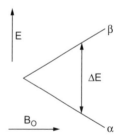

> The energy difference between an α spin state and a β spin state depends on the size of the external magnetic field B_0.

This equation relates μ, the magnetic moment of the proton, the magnetic flux density, B_0 (i.e. the applied field strength), and the frequency, ν, which has to be employed before energy can be absorbed by the proton to bring about the inversion of the magnetic moment. The value of ν necessary to satisfy Equation (4.1) clearly depends on the magnitude of the applied field. For values of B_0 typical of NMR experiments, ν is in the **radiofrequency** region of the electromagnetic spectrum. This means that the values of ΔE appropriate to NMR are much smaller than the energy differences associated with rotational, vibrational, and electronic changes discussed in Chapters 2 and 3.

The simplest way in which the NMR experiment may be carried out is to vary the magnetic field and keep the applied frequency, ν, constant in a search for the exact condition when absorption of energy leads to 'flipping' of the hydrogen atom's magnetic moment. In this type of experiment a radiofrequency oscillator is the source of

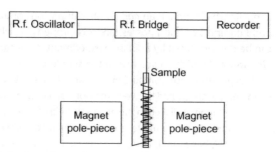

Figure 4.2 Basic features of a continuous-wave NMR spectrometer.

electromagnetic radiation of fixed frequency, v, and an electromagnet is employed to generate the variable magnetic field (see Figure 4.2).

When the magnitude of the applied field is such that the radiofrequency radiation is absorbed (at *resonance*), an imbalance is produced in a radiofrequency bridge, and the resulting signal can be amplified and fed to a recorder as a plot of absorption of energy against frequency. To achieve as sharp an absorption as possible the magnetic field must be varied very slowly. As can be seen from Equation (4.1) the resonance condition can be achieved for a fixed *ratio* of v to B. For example, for a frequency, v, of 60 MHz, the appropriate field is *ca.* 1.4 Tesla (*ca.* 14000 gauss), an arrangement which might be employed in a small 'bench-top' NMR spectrometer. For a spectrometer with v 400 MHz, typical of many modern spectrometers, the field for resonance is 9.4 Tesla (which will require a superconducting magnet). Similarly, for a spectrometer where v is 500 MHz, B_0 is 11.7 T.

In practice, signals can be readily detected for the hydrogen atoms in a small quantity *(ca.* 0.5 cm^3) of a 'neat' substrate (as a liquid) or of a substrate with hydrogen atoms in a suitable solvent which would normally have no hydrogen atoms (e.g. a deuterated solvent) and therefore no absorption in this region. Signals can be detected from substrates at low concentration *(ca.* sub-millimolar) which means that NMR spectra can be recorded even when only a few milligrams of compound are available.

In a typical hydrogen-containing sample there will be a large number of ^{1}H nuclei, distributed between the two energy levels previously described; the ratio of the number in the upper energy level (N_u or N_β) to the number in the lower energy level (N_l or N_α) is given by the Boltzmann Distribution (Equation (4.2)), where T is the Kelvin temperature and k is Boltzmann's constant.

$$N_{upper} = N_{lower}\, e^{-\Delta E/kT} \qquad (4.2)$$

This relationship is generally applicable to the statistical distribution of particles between possible energy levels, and it tells us, for example, that for an energy difference (ΔE) which is large compared with the typical thermal energy of the molecules ($\sim kT$), the lower energy level is much more highly populated as, for example, in the typical vibrational and electronic energy levels discussed in Chapters 2 and 3. However, for hydrogen atoms at room temperature in a typical magnetic field ΔE is considerably smaller than kT: the ratio N_u/N_l is nearly unity and the population of hydrogen atoms is almost equally divided between the two energy levels. For a million hydrogen atoms, there will be just a few more in the lower level than in the upper level. When irradiation of the sample with radiation at the resonance frequency (v) takes place, transitions in both upward and downward directions occur (that is, absorption and emission take place). Overall *absorption* results because of the slight excess of nuclei in the lower level.

The smaller the ratio N_u/N_l, the greater is the sensitivity (i.e. the greater the excess in the lower level); as will be seen from Equation (4.2), this means that a higher applied field (and higher v, therefore) provides a significant advantage in sensitivity whilst also providing greater *resolution* (the opportunity to separate signals with similar chemical shifts, as described in the next section).

The variable-field approach described here suffers from the limitation that it takes a relatively long time to scan the magnetic field range and it has now been replaced in the majority of spectrometers by the Fourier Transform method. This method of data measurement offers many benefits and is discussed later, in section 4.7.

4.3 ^{1}H NMR spectra of organic molecules

On the basis of the previous discussion it would be anticipated that, since NMR is a nuclear phenomenon, the resonance condition for different hydrogen atoms in a variety of molecules should not be affected by the electronic environment of each. To a certain extent this appears to be true; that is, for a fixed frequency (v), hydrogen atoms in a variety of molecules absorb energy at approximately the same value of B_0. However, minor differences of crucial importance do exist.

For example, when a low resolution NMR spectrum of ethanol, CH_3CH_2OH, is recorded it is found that three distinct resonances occur at slightly different field strengths (the field differences between the resonances are much smaller than the magnitude of the field). The NMR spectrum of ethanol is shown in Figure 4.3. The areas under the peaks are in the ratio 1:2:3, and we can conclude that the OH hydrogen atom, the two CH_2 hydrogen atoms and the three CH_3 hydrogen atoms have separate resonances.

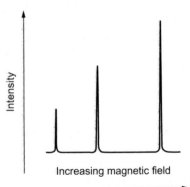

Increasing magnetic field

Figure 4.3 Low resolution ^{1}H NMR spectrum of ethanol CH_3CH_2OH.

Similar observations are made for other organic molecules. For example, the spectrum of 2-methylpropan–2–ol (t-butyl alcohol, $(CH_3)_3COH$), shows two peaks with the relative intensity ratio 9:1, and diethyl ether, $CH_3CH_2OCH_2CH_3$, exhibits a spectrum with two peaks in the ratio 3:2. In this case, the two methyl groups are in equivalent environments, but this is different from the environment of the two methylene (CH_2) groups.

Figure 4.4 shows the two-peak spectrum from ethanal (acetaldehyde, CH_3CHO), and illustrates the integration trace (upper curve). The height of each step in this trace is proportional to the area under the appropriate resonance and hence to the number

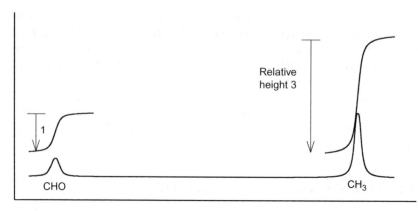

Figure 4.4 Low resolution spectrum of ethanal, CH_3CHO, showing resonances and integration trace.

of hydrogen atoms in each group. For this example, the ratio is 1:3, corresponding to the separate resonances of CHO and CH_3 hydrogen atoms, respectively.

Hydrogen atoms which undergo absorption of energy at different fields are said to have different **chemical shifts**, which depend on the environments of particular atoms in a molecule.

Chemical shifts

The observation that in CH_3CH_2OH, for example, the three types of hydrogen atom absorb at slightly different applied magnetic fields suggests that somehow each type of hydrogen nucleus does not experience exactly the same magnetic field, B_0. In practice, it can be shown that, in the presence of the applied field, small *local* magnetic fields are induced in the neighbourhood of the nucleus. Each nucleus now experiences an effective field,

$$B_{effective} = B_0 + B_{local}$$

and, since B_{local} is found to be proportional to B_0, we can write

$$B_{effective} = B_0(1+\sigma)$$

Clearly, the size and direction of σ (a measure of the chemical shift) determines the applied field that is needed to achieve the unique condition for each resonance ($h\nu = 2\mu B_{effective}$). It is known that the local fields arise from electron circulations induced by the applied field, and two rather different cases can be distinguished.

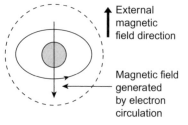

Figure 4.5 The production of local fields by induced electron-circulation.

External magnetic field direction

Magnetic field generated by electron circulation

(i) Figure 4.5 indicates the direction of electron circulation round a nucleus that is induced by the external magnetic field for a spherical electron cloud (this corresponds effectively to an electron in the 1s-orbital on the hydrogen atom). The electron motion creates an induced magnetic field which *opposes* the main field. This results in the nucleus experiencing a smaller overall field than that applied (the nucleus is said to be *shielded* from the main field) and the resonance condition must now be achieved with a higher applied field.

For CH_3CH_2OH the extent of shielding differs for each type of hydrogen atom because of different *electron densities* around the CH_3, CH_2, and OH hydrogen atoms. For the OH hydrogen atom the electron density around the nucleus will be relatively low because of the adjacent electronegative oxygen atom. However, for the CH_2 and CH_3 hydrogen atoms, which are progressively further from the oxygen atom, the electron density will be progressively greater. The CH_3 hydrogen atoms, being most highly shielded, resonate at the highest field (Figure 4.3).

In general, we can relate the magnetic fields necessary for resonance of different hydrogen atoms to the structure of a molecule and, in particular, to the electron-withdrawing properties (negative inductive effect) of the atoms present.

(ii) The second type of electron circulation which contributes to local fields in appropriate cases is that which can be induced in molecules containing double-bonds (i.e. molecules containing electrons in π-orbitals). A particularly clear example is provided by the marked effect when delocalization around an aromatic ring is possible. This is called a ring current and is illustrated for benzene in Figure 4.6.

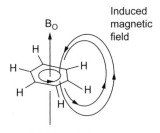

Figure 4.6 Induced electron circulation in benzene: the ring-current effect.

When the benzene molecule is oriented perpendicularly to the applied field, electron circulation is induced in the overlapping $p(\pi)$-orbitals and this leads to the production of local magnetic fields around the molecule. These local fields augment the applied field at the hydrogen atoms, so that they resonate at a *lower* external field strength. The hydrogens are said to be *deshielded*. The rapid motion of the molecules means that only for a fraction of the time is the molecule in the particular orientation depicted but nevertheless the ring-current contribution is a dominant effect and is a readily-applied criterion used to distinguish aromatic groups.

This type of effect also contributes to the observed chemical shifts for hydrogen atoms close to carbonyl, alkynic, and certain other groups.

The measurement of chemical shifts

Measurements of chemical shifts are not quoted in field units, because, as we have seen, the field difference (ΔB) between two peaks depends upon the applied field (B_0) of the spectrometer employed (Equation (4.1)).

The following procedure provides a way round this problem and leads to an acceptable universal scale. First, a suitable reference compound is chosen; tetra-methylsilane (TMS), $Si(CH_3)_4$, is widely used for ^1H spectra as it is inert and has a single resonance at a position which does not often overlap with other signals (there is only one type of hydrogen atom in the molecule). Then, the NMR spectrum of the compound under investigation, with TMS added, is recorded. There will be a field difference ΔB between the absorption of a hydrogen nucleus in the sample and the reference; $\Delta B/B_0$ gives a measure of the chemical shift which is *independent of the applied field*. The resulting numbers are very small indeed and it is more convenient to multiply them by 10^6 and to refer to the resulting measure of chemical shift, δ, in parts per million (ppm).

$$\delta/ppm = \frac{\Delta B}{B_0} \times 10^6$$

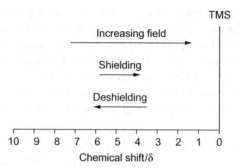

Figure 4.7 The δ-scale of chemical shifts.

This leads to a scale of chemical shifts with δ = 0 for the TMS hydrogen atoms and with most other hydrogen atoms in organic molecules in the range δ 10–0 (Figure 4.7). For example, the chemical shift of the hydrogen atoms in benzene is δ 7.25, and the shifts of hydroxyl, methylene and methyl hydrogen atoms in ethanol are approximately δ 5.2, 3.65, and 1.2 respectively.

Chemical shifts, expressed as δ-values, are characteristic of the hydrogen nuclei in the compounds concerned and are independent of the spectrometer. Further, it is quite rare to find a marked dependence of the δ-value on the solvent. Normally, chemical shifts are also independent of temperature but sometimes chemical exchange processes interfere and complex variations are observed (section 4.5). We should also remember that the chemical shift scale is independent of the combination of field and frequency used for a particular spectrometer (even though there are significant advantages in using higher fields and frequencies, as described in section 4.2).

Relationship between chemical shifts and molecular structure

The dependence of δ upon molecular structure is understandable in terms of the magnitudes of the local fields produced by the electron circulation effects described earlier. For example, although a CH_3 group in an alkane has a characteristic δ value of approximately 0.9, substitution in the molecule of a group with a negative inductive effect lowers the electron density round the hydrogen atom and hence increases the δ-value.

This increase is especially marked when the substituent is highly electronegative. Thus, along the series of halogeno-alkanes, CH_3X (X = I, Br, Cl, and F), the chemical shift of the methyl group hydrogen is progressively increased (see Table 4.1): the highest δ value for the hydrogen atoms in fluoromethane indicates that the hydrogen atoms in this molecule are the least shielded from the external field (as expected, because fluorine is the most electron-attracting substituent).

The effect becomes even more pronounced when more than one electronegative element is present; for the hydrogen atom in $CHCl_3$, for example, δ = 7.29. *Since the chemical shift of a hydrogen atom depends on its immediate environment within a molecule, the resonance positions (δ-values) of hydrogen atoms in compounds of different structural types prove diagnostically useful; some approximate values for typical groups are given in Table 4.2.*

The increase in the δ-values along the series CH_3–C, CH_3–N, and CH_3–O (in alkanes, amines and methoxy compounds (methyl ethers), respectively), again reflects the increasing negative inductive effect of the substituent group.

Table 4.1 Chemical shifts/δ of hydrogen atoms in halogenomethanes

CH_3F	4.25
CH_3Cl	3.05
CH_3Br	2.70
CH_3I	2.25

Table 4.2 Chemical shifts for hydrogen atoms in organic compounds

Group	Type of compound	Chemical shifts δ*
H_3C-C	alkane	0.9
$C-CH_2-C$	alkane	1.3
$H_3C-C=C$	alkene	1.6
ester, acid group	ester, acid	2.0
ketone group	ketone	2.1
H_3C-N	amine	2.3
H_3C-O	methyl ether	3.3
methyl ester group	methyl ester	3.7
$CH_2=C$	alkene	4.7[†]
$H-C$	arene	7.3[†]
alkanal group	alkanal	9.7[†]

*Typically ± 0.1; substitutent effects may be greater especially for compounds indicated[†]

Methyl esters of alkanoic acids, RCO_2CH_3, as might be expected, resemble methoxy compounds (methyl ethers) and have δ (CH_3) in the range 3.6–3.8. In contrast, both methyl ketones and alkyl esters of ethanoic acid (acetic acid), CH_3CO_2R, generally show a methyl group absorption at δ 2.0, this being the characteristic chemical shift of a methyl group adjacent to a carbonyl group.

Groups which have an inductive effect but which are further removed in the molecule have a decreased but sometimes noticeable effect: an example is the δ-value of 1.20 for the methyl-group hydrogen atoms in ethanol (raised from ca. 0.9 by the effect of oxygen, but not as high as the δ-value of ca. 3.3 for molecules in which a CH_3 group is directly attached to the oxygen atom).

Methylene hydrogen atoms (CH_2) have slightly higher δ-values than similarly-placed CH_3 hydrogen atoms; the characteristic δ-value for a $-CH_2-$ group in an alkyl chain is 1.3.

A hydrogen atom attached to a carbon atom in the double bond of an alkene has a higher δ-value (typically about 4.7) than hydrogen atoms in saturated analogues. This shift is associated with the sp^2, rather than the sp^3, hybridization of the alkenic carbon to which the hydrogen atom is bonded. Since few other types of hydrogen atom absorb in this area of the NMR spectrum, absorptions at δ ca. 5 have particular diagnostic value.

For benzene and other aromatic compounds the ring current effect (see Figure 4.6) explains the unusually high δ-values observed (*ca.* 7).

The lower δ-value for a hydrogen atom attached to an alkyne triple bond (*ca.* 2.0) also results from electron circulation (in the triple bond, around the molecular axis) which in this molecule provides *a shielding* effect at the hydrogen atoms. You may like to work out why the effect is opposite in direction to that observed for the hydrogen atoms in benzene.

When an unknown compound is studied with the aid of NMR spectroscopy, the positions of the absorptions (i.e. the δ values) give a fairly clear indication of the local environment of each type of hydrogen atom in the molecule, and the data given in Table 4.2 therefore prove diagnostically useful. In addition, remember that the integration trace also gives extra information of the relative numbers of hydrogen atoms in each group.

The NMR spectrum of 1,4-dimethylbenzene (*p*-xylene), shown in Figure 4.8, provides a good example of the way in which a structural analysis can be carried out. Thus, in addition to the peak at δ 0 from the added standard (TMS), two absorptions are detected: the ratio of the areas under the two absorptions is 2:3, and the peak at low-field (δ 7.05) is good evidence for the presence of aromatic-ring hydrogen atoms. The other peak, at higher field, has a δ-value of 2.29, characteristic of methyl groups attached to an arene. In this way, the structure is confirmed. Note that there are only *two* different types of hydrogen atom in the molecule and therefore two absorptions.

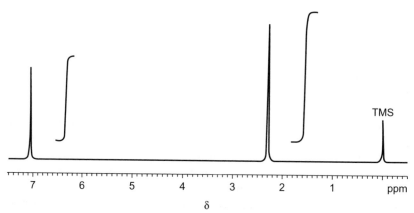

Figure 4.8 ^{1}H NMR spectrum of 1,4-dimethylbenzene, 4-$CH_3C_6H_4CH_3$.

Although the information about the number and environment of a given type of hydrogen atom in a molecule is undoubtedly of considerable assistance in structure determination, there is further helpful information which can be obtained from NMR spectra if they are recorded under conditions of higher resolution.

Spin–spin couplings (splittings)

When care is taken to ensure that the magnetic field across the sample is homogeneous, high-resolution conditions are achieved and some of the characteristic absorptions discussed previously are observed to be split into several components. We already have the basis for an explanation for this because we started out by saying that, when a simple bar magnet is placed in a magnetic field, attractive and repulsive forces result. This situation also exists when two magnetically active nuclei are adjacent to one another. Clearly this situation is common in many molecules. For example, Figure 4.9 is a high resolution spectrum of ethanal which shows that the peak of area 3 (from the CH_3 group at δ 2.21) is split into two (compare this with the low-resolution spectrum, Figure 4.4); the absorption is said to be a *doublet* and the two component lines have the same intensity (1:1). On the other hand, the absorption from the CHO hydrogen (δ 9.80) is split into a *quartet*, with relative intensities for the four lines of 1:3:3:1. The areas under the two *groups* of lines (i.e. the groups with different chemical shifts from the two types of hydrogen atom) are still in the ratio 3:1, as indicated by the integration trace.

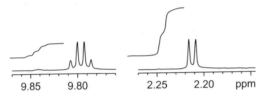

Figure 4.9 High resolution spectrum of ethanal, CH_3CHO.

The splittings can be explained by considering first the methyl-group hydrogen atoms. These experience an extra local magnetic field, which is due to the magnetic moment of the CHO hydrogen atom. This proton's magnetic moment must be aligned either *with* or *against* the applied field, so that in the CH_3–CH fragment the CH_3 hydrogen atoms experience an extra local magnetic field which can either *augment* or *oppose* the main field. Thus, the resonance condition for the CH_3 hydrogen atoms can be achieved by two frequencies which correspond to situations where the externally applied field is augmented by, or opposed by, the local field from the aldehydic hydrogen atom. For a collection of CH_3CHO molecules some of the CHO-group hydrogen atoms will augment the main field whereas in other ethanal molecules the magnetic moment of the CHO hydrogen atom will be opposed to the main field: absorptions will be seen for the CH_3 hydrogen atoms in both of these possible environments and therefore there are two peaks for the resonance from the CH_3 hydrogen atoms. The CH_3 resonance is said to be 'split' by the single hydrogen atom, and the distance between the two peaks of the doublet is referred to as the coupling (or spin–spin coupling, since the magnetic moments are effectively produced by the spinning motion of the nuclei). Because both these states are almost equally populated (section 4.2) the two resonances have the same area.

A splitting into a doublet is characteristic of a group of equivalent hydrogen atoms (CH_3 in this example) split by *one* neighbouring hydrogen atom. We can now derive the splitting patterns for larger numbers of neighbouring hydrogen atoms, as, for example, when the resonance from a hydrogen atom is split by an adjacent CH_2 or CH_3 group.

Splitting from two hydrogen atoms. Consider the spectrum from diethyl ether, $CH_3CH_2OCH_2CH_3$ (Figure 4.10) which has two main absorptions, from the hydrogen atoms of the CH_2 and CH_3 groups. The two sets of CH_3 hydrogen atoms (δ 1.13) will each experience local fields from adjacent –CH_2– hydrogens, and we must therefore

The CH_3 protons are in resonance: there are two possible magnetic fields generated by the CH proton which can reinforce or reduce the main field thus there are two possible values of the applied field which, when combined with the two possible local fields, give the *overall* field for resonance. These two situations result in the resonance of the CH_3 protons being split into a **doublet**.

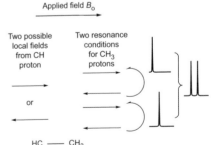

There are four ways of arranging the two nuclear spins of a CH_2 group. These produce three local magnetic fields which split the resonance from the adjacent hydrogen atoms into a **triplet**.

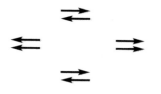

There are eight ways of arranging the three nuclear spins of a CH_3 group. These produce four local magnetic fields which split the resonance from the adjacent hydrogen atoms into a **quartet**.

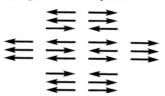

work out the possible 'arrangements' of the latter. Two CH_2 hydrogen nuclei may be both aligned with their magnetic moments in one direction (which we can represent ⇉), both in the other direction (⇇), or in opposite directions (⇄ or ⇆). This gives three possible local magnetic fields, of which, for a collection of molecules, the arrangement with the magnets opposed (⇆ and ⇄) can be achieved in two possible ways and will be twice as common as the others. Thus the splitting of the CH_3 resonance produces a 1:2:1 pattern, a **triplet**.

Splitting from three hydrogen atoms. In the two previous examples (CH_3CHO and $CH_3CH_2OCH_2CH_3$) to fully assign the spectrum we have to realize that the resonance signal from the CHO hydrogen atom in CH_3CHO must be split by the CH_3 group and that the two sets of CH_2 hydrogen atoms in $CH_3CH_2OCH_2CH_3$ (at δ 3.4) must be split by the effect of the adjacent CH_3 hydrogen atoms. We therefore have to work out the possible local magnetic fields provided by the three magnetic moments of a CH_3 group. These are shown to the left.

There are four possible resultant magnetic fields, two of which can be achieved in three possible ways: thus a CH_3 group splits the resonance of a neighbouring nucleus into a four-line pattern (quartet) with relative intensities 1:3:3:1, as confirmed by the splitting of the CH_2 absorption in Figure 4.10 at (δ 3.4) and of the CHO absorption in Figure 4.9 (δ 9.8).

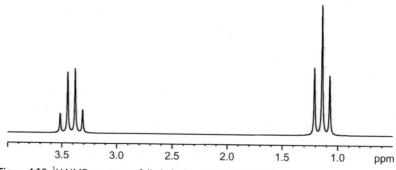

Figure 4.10 ^{1}H NMR spectrum of diethyl ether, $CH_3CH_2OCH_2CH_3$.

More than three hydrogen atoms. In the fragment $-CH_2-CH-CH_2-$, in which the two CH_2 groups are equivalent (and have the same chemical shift), the single hydrogen atom will interact with four equivalent hydrogen atoms; the resulting pattern is a 1:4:6:4:1 **quintet** (as a self-test you might like to work out the arrangement of spins and multiplicities which give this pattern). Five equivalent neighbouring hydrogen atoms produce a splitting pattern of 1:5:10:10:5:1, and six equivalent hydrogen atoms (e.g. in a $(CH_3)_2CH-$ group) split the single hydrogen atom's resonance into a 1:6:15:20:15:6:1 pattern. The splitting patterns from numbers of equivalent neighbouring protons follow the coefficients of the terms in the binomial expansion and are conveniently expressed in the form of 'Pascal's Pyramid' (triangle), as shown on the next page. Note that you can generate this table of patterns from the starting point given above (i.e. 1:1, 1:2:1, 1:3:3:1) by writing a new 'line' with each extra number being the sum of the two above it.

Number of hydrogen atoms causing splitting	Splitting pattern produced (relative intensities of lines)
1	1 1
2	1 2 1
3	1 3 3 1
4	1 4 6 4 1
5	1 5 10 10 5 1
6	1 6 15 20 15 6 1

Recap: Use of shifts, integrations, and splittings in analysis

Before proceeding further, it is perhaps worthwhile reminding ourselves of the various stages in the analysis of a complex NMR spectrum by reference to that of CH_3CHO.

First, under low resolution conditions there are two absorptions, of relative area 3:1, (measured from the integrations), from the CH_3 and the CHO hydrogens, respectively. These different groups have different *chemical shifts* (δ-values) which are typical of the chemical environment of the two types of hydrogen atom (Table 4.2).

Second, under high resolution conditions, *splittings* can be seen. Interaction of the CH_3 hydrogen atoms with the neighbouring CHO hydrogen atom means that the former resonance is a 1:1 doublet; interaction of the neighbouring CHO hydrogen atom with the CH_3 hydrogen atoms means that the aldehydic hydrogen's resonance is a 1:3:3:1 quartet (see Figure 4.9).

A little more information on splittings

The size of the coupling (see Figure 4.9)—i.e. the separation between the peaks of each multiplet—is a measure of the energy of the magnetic interaction between the two types of proton (CH, CH_3), and is the same for both resonances. The coupling referred to as J_{HH} is customarily quoted in frequency units ($\Delta v/Hz$); it is, for a given compound, independent of the magnitudes of the characteristic radiofrequency and applied magnetic field of the NMR spectrometer employed.

A word of explanation is necessary here since the measurement involves (usually) conversion from the δ-scale: the problem is to express a separation (measured as $\Delta\delta$) as Δv (Hz). From the definition of δ we have:

$$\Delta\delta = \frac{\Delta B.10^6}{B_0}$$

and, from Equation (4.1)

$$\frac{\Delta B}{B_0} = \frac{\Delta v}{v}$$

so that

$$\Delta v = \frac{v.\Delta\delta}{10^6}$$

In the example CH_3CHO, the separation between the lines (for both the 1:1 and the 1:3:3:1 patterns) is 0.0075 δ on a 400 MHz spectrometer (such as that used for the

With a 400 MHz spectrometer (i.e. $v =$ 400×10^6 Hz), such as that used for many of the ^{1}H NMR spectra shown in this book, 1δ unit is equivalent to 400 Hz. For a 100 MHz spectrometer, 1δ unit is equivalent to 100 Hz.

majority of figures in this chapter). This corresponds to 3 Hz and is always 3 Hz, no matter which spectrometer is employed. The separation *measured as* $\Delta\delta$ is 0.03 on a 100 MHz spectrometer. It is referred to as $^3J_{HH}$, i.e. the splitting between two hydrogens, three bonds apart. Examples of information which can be derived from the magnitude of the splittings will be provided in the next section.

Analysis of the splitting patterns in the ^{1}H NMR spectra of organic compounds is usually fairly straightforward because appreciable couplings normally occur only between hydrogen atoms on *neighbouring* atoms (information about the alignment of one magnetic moment is transmitted to other nuclei through the bonds and the effect dies off rapidly with the number of intervening bonds). This makes the technique extremely effective for distinguishing isomeric alkyl structures (straight-chain and branched): see, for example, the worked examples in section 4.4.

It must also be remembered that hydrogen atoms in exactly equivalent environments in a molecule do not split each other. Thus, for CH_3CHO, for example, the methyl hydrogen atoms, while split by the CHO hydrogen, do not split each other. This is because they are all identical and in resonance together (at the same chemical shift and providing no fixed local field).

Also note that in this example, as in many others, the splitting pattern observed matches the expected 1:1 and 1:3:3:1 relative intensities: Sometimes, the peaks within these groups are slightly larger in the direction of the resonance of the group responsible for the splitting (see e.g. Figures 4.10, 4.11, and 4.16). The distortion (which can be ignored here) becomes more pronounced when chemical-shift differences become small, though analysis is usually still possible.

4.4 Examples of spectra showing spin–spin splittings

(a) Butanone (methyl ethyl ketone), $CH_3COCH_2CH_3$

The ^{1}H NMR spectrum (Figure 4.11) contains a single peak from the methyl group adjacent to the carbonyl group, with the expected chemical shift (δ 2.15; see Table 4.2). This peak is not split because there are no hydrogen atoms on the adjacent carbon atom.

The other peaks are from the hydrogen atoms in the CH_2 group (δ 2.55) and the other CH_3 group (δ 1.05); the CH_2 peak has the higher δ-value because of the effect of the adjacent carbonyl group and the resonance appears as a 1:3:3:1 quartet because of the methylene group's interaction with the CH_3 group. The CH_3 resonance is split into

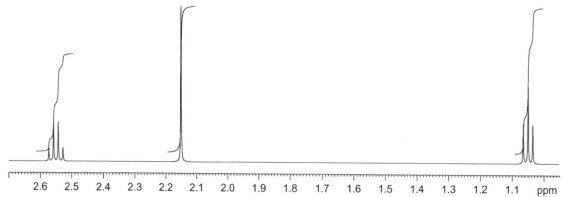

Figure 4.11 ^{1}H NMR spectrum of butanone, $CH_3COCH_2CH_3$.

a 1:2:1 triplet by the CH_2 hydrogen atoms (these two multiplets form the characteristic pattern seen for an *ethyl* group).

As can be seen from Figure 4.11, the integration trace indicates the relative numbers of hydrogen atoms in the groups with different chemical shifts.

(b) I,3-Dibromopropane, $BrCH_2CH_2CH_2Br$

The NMR spectrum, shown in Figure 4.12, consists of two groups of resonances, the numbers of hydrogen atoms concerned being 2:1 (i.e. the relative total areas, as indicated by the height of superimposed steps in the integrated curve for each resonance). These are evidently the four outside and the two central methylene hydrogen atoms, respectively, with the expected δ-values; the outside methylene groups absorb at higher δ value because of the inductive effect of the bromine atoms.

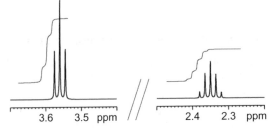

Figure 4.12 ^{1}H NMR spectrum of I,3-dibromopropane, $BrCH_2CH_2CH_2Br$.

The central CH_2 resonance is split into a 1:4:6:4:1 quintet by the *four* neighbouring hydrogen atoms; the outside CH_2 groups each have one neighbouring CH_2 group and hence appear as 1:2:1 triplets.

(c) Bis-(I-methylethyl) ether (di-isopropyl ether), $(CH_3)_2CHOCH(CH_3)_2$

The spectrum (Figure 4.13) indicates the presence of two different types of hydrogen atom, the numbers of each type being in the ratio 1:6. The methyl group resonance (at δ 1.1) is split into a doublet, since each CH_3 group hydrogen atom experiences an interaction with the neighbouring single hydrogen atom. If you look carefully, you will see that the CH resonance (δ 3.6) is split into a septet, characteristic of interaction with six equivalent hydrogen atoms.

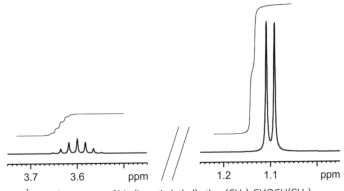

Figure 4.13 ^{1}H NMR spectrum of bis-(I-methylethyl) ether $(CH_3)_2CHOCH(CH_3)_2$.

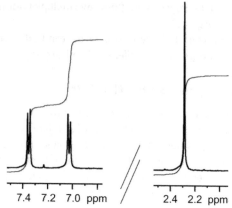

Figure 4.14 ^{1}H NMR spectrum of 1-methyl-4-bromobenzene (p-bromotoluene, 4-$CH_3C_6H_4Br$).

(d) 1-Methyl-4-bromobenzene (*p*-bromotoluene), 4-$CH_3C_6H_4Br$

The spectrum (Figure 4.14), recorded for a solution in CCl_4 shows clearly the resonances from aromatic (δ 7.4–7.0) and aliphatic (δ 2.3) hydrogen atoms, in the ratio
4:3. The methyl group resonance is not visibly split (because there are no adjacent
hydrogens that yield a resolvable coupling) but the aromatic hydrogens appear as
two non-equivalent pairs: H_A, δ 7.35, and H_B, δ 7.02. The signals for H_A and H_B are
both doublets because of the splitting for each hydrogen atom by the adjacent non-
equivalent hydrogen atom (J_{HH} 9 Hz). This four-line pattern is typical of a 1,4-disubsti-
tuted benzene ring with two different substituents.

For *p*-bromotoluene there are two pairs
of chemically equivalent protons that are
magnetically inequivalent.

Magnitudes of splittings

Not only does the observation of the splitting patterns in NMR spectra indicate which
groups are attached to others in an organic molecule, but also, in certain cases, it is
found that the magnitude of the splitting (J) is informative. The splitting between two
hydrogen atoms in adjacent C–H bonds is known to reflect an interaction transmitted through the electrons in the bonds and, in general, is found to be approximately
proportional to $\cos^2\theta$ (where θ is the dihedral angle between the two C–H bonds). The
following example (structures 4.1, 4.2) shows the difference between the interaction
between two adjacent C–H bonds in a cyclohexane ring, i.e. between the equatorial
and/or axial C–H bonds, which emphasizes the use of such splittings in determining
conformational geometry (note that the full 3-dimensional shape (conformation) of
the cyclohexane ring is illustrated on page 149, section 6.7).

J_{HH} 10–12 Hz (θ 180°)   (4.1)

J_{HH} 2–5 Hz (θ 60°)   (4.2)

There are also important differences between the interactions in alkenes in which
the hydrogen atoms are not all equivalent (if they were, they would all have the same
δ-value and hence have no observable splitting) and the following splittings, given in
the margin, are typical.

The recognition of these differences sometimes enables a choice to be made between various possible isomeric structures. For example, in one of the isomeric forms of 3-phenylpropenoic acid (cinnamic acid, $C_6H_5CH=CHCO_2H$) the two alkene protons resonate at δ 7.83 and 6.46, with a coupling (J_{HH}) of 17 Hz: this must therefore be the *trans* isomer. Similarly, the splitting between two hydrogen atoms next to each other in a benzene ring is typically 7 Hz (similar to the *cis* alkene here).

Figure 4.15, the rather more complicated ^1H NMR spectrum from phenylethene (styrene), provides an example of an alkene with *three* non-equivalent hydrogens (H_A, H_B, H_C), marked up to illustrate the three different J_{HH} splittings, with magnitudes in the ranges noted above. Note that where *one* hydrogen atom's NMR absorption is split by interaction with *two* non-equivalent protons a *doublet* of *doublets* results. Note also the complexity of the aromatic region, δ 7.3–7.5, which is referred to again in Worked example 4.2, page 85.

Geminal: 1–3 Hz Cis: 8–10 Hz Trans: 14–17 Hz

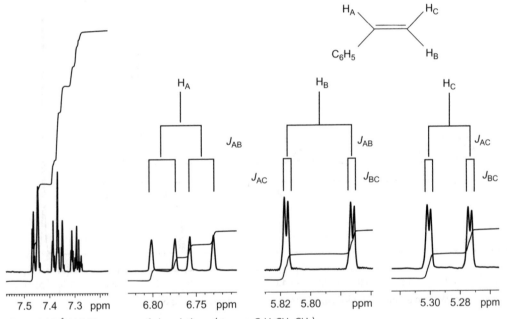

Figure 4.15 ^1H NMR spectrum of phenylethene (styrene, $C_6H_5CH=CH_2$).

In some spectra, spin–spin couplings and even resonances which you might expect to see are absent. This often happens when protons (such as OH) are exchangeable with those in the solvent. The observation of such behaviour is indicative of chemical exchange and is discussed further in section 4.5.

Worked examples

At this stage you are encouraged to attempt to assign a structure that fits each of the spectra shown in Figures 4.16–4.18, for each of which the molecular formula is given.

These are followed by a discussion which gives the answers and some brief explanatory notes. Remember that the spectra offer three pieces of vital information: the chemical shifts (related to the types of hydrogen atom present), the integration traces (the heights of the steps are proportional to the relative numbers of hydrogen atoms

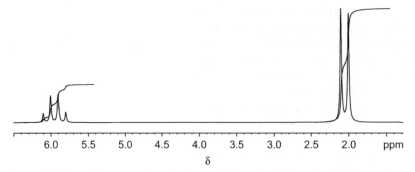

Figure 4.16 ^{1}H NMR spectrum of Worked example 4.1; molecular formula $C_2H_4Cl_2$.

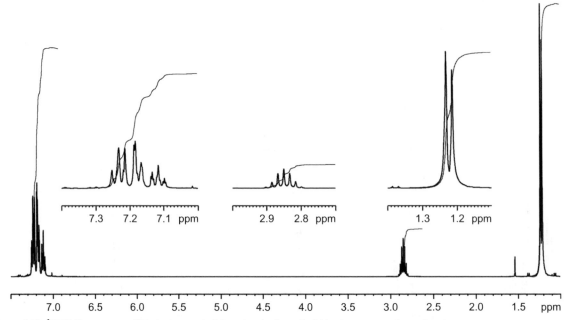

Figure 4.17 ^{1}H NMR spectrum of Worked example 4.2; molecular formula C_9H_{12}. Inset, expansions of peaks.

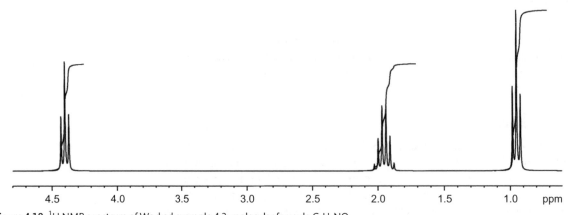

Figure 4.18 ^{1}H NMR spectrum of Worked example 4.3; molecular formula $C_3H_7NO_2$.

in each group), and the splittings (which depend on the number of hydrogen atoms in immediately adjacent groups).

Discussion

4.1   Figure 4.16 is the spectrum of 1,1-dichloroethane, CH_3CHCl_2. There are clearly two types of hydrogen atom in the molecule, the number(s) in each group being in the ratio of the integrated intensities (3:1). This strongly suggests the presence of CH_3 and CH groups, which is confirmed by the splittings of 1:3:3:1 (of the CH absorption due to interaction with three equivalent hydrogen atoms) and 1:1 (the doublet from splitting of the methyl group absorption by the single hydrogen atom). The high δ-value (5.9) for the single hydrogen atom reflects the negative inductive effect of the chlorine atoms.

4.2   Figure 4.17 is the spectrum of an aromatic compound, as judged from the characteristic absorption at δ 7.2. The aliphatic part of the molecule (δ 1.2 and 2.8) has two types of hydrogen atom, apparently in the ratio 1:6 (as deduced from the integrated trace). The high-field peak is typical of a C–CH_3 group, so that the part-structure–$CH(CH_3)_2$ may be suggested. This is confirmed by the splittings; the methyl hydrogen absorption is split into a doublet by interaction with one hydrogen (CH) and the absorption of the single hydrogen is split by interaction with six equivalent hydrogen atoms to give a septet (1:6:15:20:15:6:1). Since the ratio of the number of aromatic hydrogen atoms to aliphatic hydrogen atoms (measured from the integrations) is approximately 5:7 the structure must be (1-methylethyl) benzene (also called isopropylbenzene and cumene), $C_6H_5CHMe_2$.

An expansion trace showing the aromatic proton region for (1-methylethyl) benzene is also included in Figure 4.17. The five aromatic protons of this molecule correspond to two chemically distinct pairs and a single proton. A fairly complex and somewhat distorted pattern results, because the chemical shift differences between the associated absorptions are small (see Exercise 4.5, page 101 and also Figure 4.15).

4.3   From Figure 4.18 we can conclude that there is a propyl group ($CH_3CH_2CH_2$–) present. Thus the high-field resonance at δ 0.95 (of relative intensity probably 3) is typical of a methyl group in an alkane, and there are two other absorptions with relative intensity 2: of these, the low-field (CH_2) peak is evidently split by a CH_2 group (to give a 1:2:1 triplet), and the splitting of the middle multiplet (evidently another CH_2) by CH_3 and CH_2 gives a 1:5:10:10:5:1 pattern. This is the spectrum of $CH_3CH_2CH_2NO_2$, although from the evidence so far presented the alternative structure $CH_3CH_2CH_2ONO$ cannot be ruled out. Distinction between these two would be made on the basis of the expected δ-values (e.g. from tables of data) for hydrogen atoms in alkyl nitrites (–CH_2ONO) and nitroalkanes (–CH_2NO_2), from other spectroscopic data (e.g. UV, IR) or from chemical evidence.

Figures 4.16–4.18 clearly demonstrate the effectiveness of ^1H NMR spectroscopy in distinguishing isomeric alkyl groups, e.g. $CH_2CH_2CH_3$ and $CH(CH_3)_2$.

4.5 Other structural information from ^1H NMR studies

(a) Hydrogen-bonding

For a solution of phenol (C_6H_5OH) in tetrachloromethane (carbon tetrachloride), the chemical shift (δ) of the hydroxyl hydrogen atom's absorption depends markedly on the concentration. We can interpret this observation in terms of the environment of the hydrogen atom concerned; at low concentrations of phenol, each molecule will be surrounded by CCl_4 molecules which will not interact appreciably with the hydroxyl group, but at higher concentrations of phenol, intermolecular association of these molecules occurs *via* hydrogen bonding. This involves an attraction between a phenol molecule's oxygen atom (which is electronegative) and the positively polarized hydrogen atom in the hydroxyl group of another phenol, molecule. By contrast, if an *intra*-molecular hydrogen bond is possible, then the –OH hydrogen atom is much less susceptible to changes in its external environment brought about by changing the concentration. This is true, for example, for 2-nitrophenol (4.3), where an internal hydrogen bond is formed between the hydrogen atom on the phenolic oxygen atom and one of the electronegative oxygen atoms of the nitro-group. For this molecule, δ (OH) is much less sensitive to changes in concentration. In some molecules the corresponding resonance positions depend on the pH of the solution, a property which is employed in MRI to differentiate tissue type and aid in the diagnosis of disease (see page 98, section 4.9).

(b) Keto–enol tautomerism: demonstration and quantitative estimation

It is also possible to use an NMR spectrometer to determine relative amounts of several constituents in a mixture. An interesting example is provided by the spectrum (Figure 4.19) recorded for a sample of pure pentane-2,4-dione (acetylacetone, $CH_3COCH_2COCH_3$) in deuterated chloroform. The spectrum confirms that the compound exists in two forms, and careful analysis indicates that these are the keto- and enol-tautomers, (4.4) and (4.5), respectively (the phenomenon is known as **tautomerism** as described in Chapter 3).

The *enol* form (4.5) is recognized by the typical alkene hydrogen absorption at δ 5.5; the broad peak at δ 15.4 characterizes a hydroxyl-hydrogen atom (the high δ-value reflects the effect of both oxygen-atoms: the hydrogen atom is hydrogen-bonded to the oxygen atom of the carbonyl group), and it disappears on shaking the compound with D_2O). The peak at δ 3.55 is characteristic of the CH_2 group between two carbonyl groups in the *keto* form (4.4). Comparison of the integrations of the δ 1.95 and

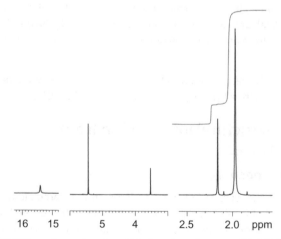

Figure 4.19 ^{1}H NMR spectrum of pentane-2,4-dione (acetylacetone, $CH_3COCH_2COCH_3$).

2.15 peaks leads to a calculated ratio of 4.75:1 for the relative amounts of enol and keto tautomers (83% enol). The dependence of the ratio of the tautomers on the nature of the solvent, and on the temperature, can readily be investigated (see also Chapter 3, page 63 and Exercise 3.2).

This type of information is not readily determined without the use of spectroscopic techniques; two tautomers are usually very easily interconvertible, so that it is not often possible to isolate and estimate the percentage of one of the individual forms.

(c) Dynamic effects

The addition of a trace of acid to a pure alcohol leads to an increased rate of hydroxyl-hydrogen atom exchange and hence to the collapse of the splitting between OH and the hydrogen atoms on the carbon atom adjacent to the oxygen atom in the alcohol. The observation of this type of *dynamic* effect in NMR spectra is not limited to examples of rapid chemical reactions (like hydrogen-atom exchange) but can also occur for some rapid intra-molecular processes.

For example, consider the NMR spectrum of N,N-dimethyl-methanamide (N,N-dimethylformamide), (4.6), at room temperature (Figure 4.20a). As indicated by their different chemical shifts the two methyl groups are not equivalent (i.e. not in identical environments). We can conclude that this arises from restriction of rotation of the –N(CH$_3$)$_2$ group about the C–N bond (which has partial double-bond character):

$$\begin{array}{c}
CH_3 \\
\diagdown \\
:N-C \\
\diagup \quad \diagdown \\
CH_3 \qquad H
\end{array}
\overset{O}{\diagup}
\longleftrightarrow
\begin{array}{c}
CH_3 \\
\diagdown \\
N{=}C \\
\diagup \quad \diagdown \\
CH_3 \qquad H
\end{array}
\overset{O^-}{\diagup}$$

(4.6)

At higher temperatures, the molecules possess more energy, the rate of the rotation about the C–N bond increases, and the methyl groups eventually appear to be

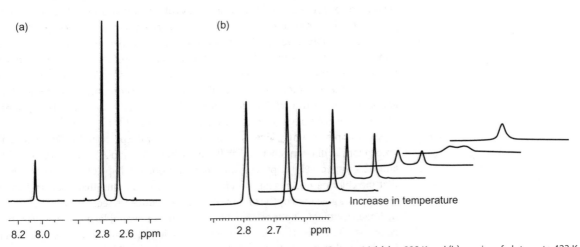

Increase in temperature

(a) 8.2 8.0   2.8 2.6 ppm

(b) 2.8   2.7   ppm

Figure 4.20 ^{1}H NMR spectrum of N,N-dimethylmethanamide (N,N-dimethylformamide) (a) at 298 K and (b) a series of plots up to 423 K showing how the influence of the dynamic effect on the line shape changes with temperature.

equivalent (that is, there is a single peak; Figure 4.20b shows how the resulting NMR spectra change with temperature).

The critical rate of rotation, at which the two separate peaks just become coalesced to a single peak, occurs when the rate of rotation is approximately equal to the difference between the two separate absorptions expressed in frequency units (i.e. Δv/Hz).

In this case $\Delta\delta$ is 0.13, which must be converted into Δv/Hz. Since the spectrometer operates at 500 MHz, Δv is 0.13×500 (see page 79, section 4.3), i.e. about 70 Hz. Thus, when the lines coalesce, at approximately 120°, the rate of rotation is of the order of 70 Hz (i.e. 70 times per second). Above this temperature, and hence at faster rates, only a single *averaged* line results for the two methyl groups.

In this example, the two groups whose positions in the NMR spectrum are being interconverted differ in their δ-values. Exactly the same arguments apply when a *splitting* (J_{HH}) disappears because of rapid exchange. Thus, in pure ethanol the OH hydrogen atom has a splitting of 5 Hz from the CH_2 hydrogen atoms (a 1:2:1 pattern). When a trace of acid is added, the hydroxyl hydrogen atom undergoes acid-catalysed exchange: as the exchange rate increases, the splitting will disappear and the coalescence point will correspond to an exchange rate of *ca.*5 Hz (five times per second). At faster rates of exchange, a single line with no splitting is observed.

Detailed analysis of the shape of the signal in this and similar cases can be employed to measure the rate of exchange (or rotation) at different temperatures. Then, an Arrhenius plot leads to an estimate of the activation enthalpy (energy barrier) for the process concerned. For *N,N*-dimethylmethanamide, the barrier to rotation is estimated as 30 kJ mol^{-1}.

In summary, NMR spectra at lower temperatures, or in the absence of reagents causing rapid exchange, can reveal details of structure which are otherwise 'averaged out' by, for example molecular motion. This can often provide extra valuable structural information, for example about preferred conformations.

4.6 NMR from other nuclei

As well as ^{1}H, other nuclei have nuclear magnetic moments; these include ^{2}H (deuterium), ^{11}B, ^{13}C, ^{14}N, ^{19}F, ^{31}P, ^{35}Cl, ^{37}Cl, ^{79}Br, and ^{81}Br.

These nuclei may give rise to *splittings* in the ^{1}H NMR spectra of appropriate molecules. However, these splittings are not always observable and, in particular, it is rarely possible to observe splittings from ^{14}N or the chlorine and bromine isotopes.

It is also possible to design NMR spectrometers which operate at different combinations of frequency, v, and field, B, than those used for ^{1}H resonance in order to detect resonance from the nuclei listed above. For example, absorption from ^{19}F atoms in fluorine-containing molecules can be detected for a spectrometer with $v = 400$ MHz at a value of B_0 of 10 T (*cf.* 9.4 T for protons). In this way, the scope of NMR spectroscopy for proving molecular structure has been considerably extended.

For example, you may like to interpret the ^{19}F NMR spectrum for ClF_3 (two main absorptions at different chemical shifts, in the intensity ratio 2:1, with splittings of 1:1 and 1:2:1 respectively) to provide information about the shape of the molecule. Note that the chlorine nucleus has no effect on the spectrum and that fluorine–fluorine splittings are governed by the same rules as those deduced in section 4.2 for hydrogen–hydrogen splittings. The spectrum rules out planar or pyramidal molecules with three equivalent fluorines. The spectra support a T-shaped geometry with two equivalent fluorines and a separate fluorine (n.b. as a result of interactions involving the two lone pairs of electrons on the chlorine atom).

The ^{11}B NMR spectrum of diborane (B_2H_6) (4.7) shows a single resonance with a large 1:2:1 splitting (implying an interaction of the boron nuclei with two hydrogen atoms) and a further 1:2:1 splitting (implying a weaker interaction of the boron atoms with two more hydrogen atoms). The two boron atoms in B_2H_6 must have identical environments within the molecule (since there is only one boron atom chemical shift), and must each interact with two different pairs of hydrogen atoms. Thus the NMR spectrum provides evidence to support the assignment of the following symmetrical bridged-structure to the molecule (electron diffraction, described in Chapter 6, allows the bond lengths and angles to be accurately determined):

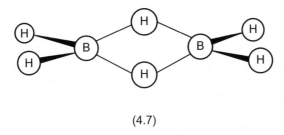

(4.7)

The ^{1}H NMR spectrum of B_2H_6 shows two resonances in the ratio 2:1 indicating, as expected, that there are two types of hydrogen (the terminal (end) and the bridge atoms, respectively).

It is particularly desirable to be able to record NMR signals from the very small proportion of carbon nuclei (^{13}C) which possess magnetic moments (you will recall that ^{12}C does not have a magnetic moment). This would enable information to be gained directly about the carbon skeleton in a molecule and confirmation to be obtained about the presence of, for example, CO or CN groups.

Many problems initially thwarted the successful detection of ^{13}C nuclei: in particular, the inherent sensitivity for detection of carbon atoms is a factor of 10^4 less than that for protons (partly because the natural abundance of ^{13}C is only 1.1%) and, in addition, the very wide spread of chemical shifts for ^{13}C in typical organic molecules (*ca.* 300 ppm) means that a considerable scan time is needed for field-swept (continuous-wave) spectrometers. Furthermore, the occurrence of complex proton-splitting patterns complicates the signals. How these problems can be (and have been) surmounted is described in the next section.

4.7 Pulsed NMR spectrometers

We have already seen that NMR spectroscopy can suffer from the limitation that detected signal strengths are very weak because the corresponding population differences are very small. While proton nuclei are the most readily detected, ^{13}C signals are very much harder to detect.

Fortunately, these problems have been very effectively solved by the design of pulsed spectrometers that are now available for the detection of all NMR active nuclei. These machines have revolutionized the scope of NMR, especially by allowing ready access to nuclei such as ^{13}C.

These instruments work by supplying for a given nucleus (say ^{13}C) a *fixed* magnetic field but a range of different **frequencies** instantaneously to the sample. This is achieved by irradiating the sample with a short **pulse** of electromagnetic radiation such that the resulting small range of applied frequencies excite the region expected to contain the resonances in the selected sample. For example, if the spectrometer employed to record a ^{13}C NMR spectrum has a magnetic field of 9.4 Tesla (which corresponds to a ^{1}H resonance frequency of 400 MHz), the frequency of the pulse (i.e. the appropriate resonance frequency for ^{13}C in this field) is 100 MHz and the spread of frequencies about this value (the bandwidth, Δv) is typically ca. 30 kHz.

The effect of the pulse of radiation is to excite all the ^{13}C nuclei whose resonance frequency lies within the range Δv. After the pulse, these ^{13}C nuclei return to their equilibrium state within a few seconds but during that time a decaying signal can be detected which contains information about all the frequencies that have been excited (i.e. the resonance frequencies). This is referred to as the Free Induction Decay (or FID: see page 18). The measured analogue signal is converted into a digital signal and then stored in the computer memory. Repeating the pulse-additive-data-storage process allows the detected signals to be added together to increase the signal strength relative to the background noise level before conversion into the NMR spectrum. Because this process enhances the signal-noise-ratio, it improves the quality of the final spectrum and thereby allows the measurement of very weak signals—including, and especially, those from ^{13}C nuclei in a sample. This approach is now used routinely to record all ^{1}H NMR spectra as well as spectra from other nuclei.

Next, a technique referred to as the Fourier Transform method (as described in Chapter 2) is employed to convert the signal stored on the computer into the final spectrum. Put simply, this process separates the complex mixture of frequencies detected after the pulse into its component parts. This can be envisaged in a similar way to the action of white light impacting on a prism because when the light leaves the prism all the colours and hues of the rainbow can be seen. In other words, the Fourier Transform process acts like a prism and separates the original signal into its component parts. The resulting spectrum is plotted in exactly the same way as described earlier, i.e. in terms of chemical shifts on the δ scale, relative to a standard, normally tetramethylsilane. An example of a Fourier Transform NMR spectrum, from ethanol, is shown in Figure 4.21. At the top, on the left (a), you see the 'beating' frequencies of the combined signals from the three different protons (the Free Induction Decay). Underneath ((b)–(d)) are the separate frequencies from each hydrogen which are expressed in ppm (δ) on the right, and, at the top, combined to give the ^{1}H NMR spectrum from ethanol (cf. page 71, Figure 4.3). At high resolution, spin–spin splittings become resolved, to give spectra identical in appearance to those described earlier.

Other nuclei

We have already seen (section 4.6) that, besides ^{1}H, other nuclei have nuclear magnetic moments: these include ^{2}H (deuterium), ^{11}B, ^{13}C, ^{14}N, ^{19}F, ^{31}P, ^{35}Cl, ^{37}Cl, ^{79}Br, and ^{81}Br. These nuclei also interact with a magnetic field and can therefore exhibit their own resonance condition as noted above, given the appropriate combinations of B and v.

Because the values of μ differ substantially from one nucleus to the next, the resonance frequencies for each nucleus in a fixed magnetic field are normally well separated. This is illustrated in Figure 4.22 and confirms that for example, the resonance frequency of ^{19}F atoms in fluorine-containing molecules in a given field can be detected at a frequency which is 94% of the ^{1}H frequency. By tuning the circuit of an NMR probe for resonance it is possible to affect, and detect, only selected values of v, in other words, probe nuclei selectively. This process is similar to tuning into different radio or

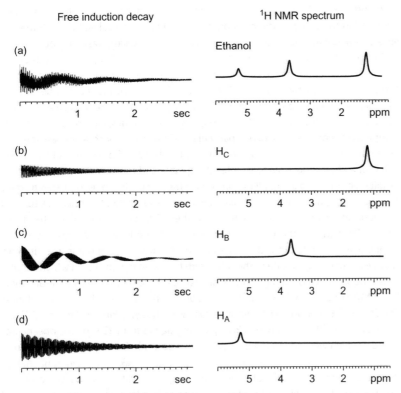

Figure 4.21 Fourier Transform ^1H NMR spectrum from ethanol (low resolution). (a) with components, (b)–(d) from the separate protons.

TV stations: it means that several NMR experiments (with different v) can be completed on the same sample in the fixed magnetic field so that several different NMR active nuclei in a sample can be monitored (e.g. ^1H, ^{13}C, ^{19}F, or ^{31}P). It is also possible to simultaneously tune the NMR probe to more than one frequency. This approach enables the user to remove the effect of one type of nuclei whilst still seeing another and is described in the next section.

Although successful NMR experiments usually require that the molecules are tumbling rapidly (i.e. in fluid solution) it is sometimes also possible to spin rapidly a sample tube containing a solid, to achieve high resolution spectra (referred to as 'magic angle' spinning).

Figure 4.22 Relative resonance frequency separations for a ^1H frequency of 400 MHz.

^{13}C NMR spectroscopy

As we have described earlier, it is particularly desirable to be able to record NMR signals from the very small proportion of carbon nuclei (^{13}C) which possess magnetic moments (you will recall that ^{12}C does not have a magnetic moment). This enables information to be gained directly about a molecule's carbon skeleton, with the presence of functional groups such as CO or CN being particularly easy to spot.

Carbon-13 spectra do not normally exhibit $^{13}C-^{13}C$ splittings because there is a very low natural abundance of ^{13}C, and the percentage of molecules which possess two adjacent ^{13}C atoms is very small. However, there will be interactions between the ^{13}C nuclei and the magnetic moments of nearby protons in the molecule, which will cause splittings (exactly as for $^1H-^1H$ coupling). The resulting patterns are often quite complicated, however, because a particular ^{13}C nucleus may show observable splittings not only from protons attached directly to it but also from protons further away.

Fortunately, it is possible to simplify the spectra and remove *all* these splittings, by simultaneous *decoupling* of all the protons in the sample in a pulsed FT spectrometer. To achieve this, the sample is irradiated with an extra range of frequencies such that all the proton resonances are excited while the ^{13}C signals are recorded. The absorption of this extra radiation means that each proton no longer provides a static local magnetic field to interact with the neighbouring carbon nuclei but instead it rapidly changes from one spin state to the other with the result that the effect which normally cause the splittings is averaged and lost. Figure 4.23 shows a typical fully-decoupled ^{13}C NMR spectrum which has been recorded in this way: this is for butanone, $CH_3COCH_2CH_3$ (whose ^1H NMR spectrum was shown earlier in Figure 4.11). This spectrum was obtained very quickly by adding signals from 32 separate observations (pulses), which were separated by a period of 3 seconds. Each of the carbon atoms in butanone gives an absorption which is a single sharp peak, with a different chemical shift; it is the range of shifts obtained, plus the simplification gained by decoupling, that is particularly useful in the study of large molecules such as proteins, alkaloids, and steroids.

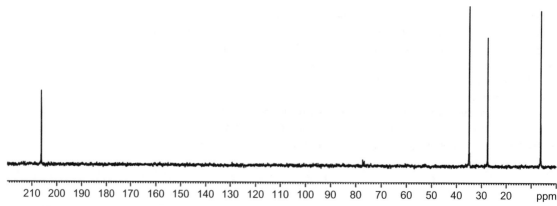

Figure 4.23 ^{13}C NMR spectrum (fully decoupled) of butanone, $CH_3COCH_2CH_3$.

For example, the steroid cholesterol acetate, whose structure is shown in (4.8), has a broad, complex and relatively uninformative proton NMR spectrum (the different alkyl fragments have very similar chemical shifts and splittings from adjacent protons, causing considerable overlap). In contrast, the fully decoupled ^{13}C spectrum shown in Figure 4.24 shows significant differences in the carbon chemical shifts, and resonances from all of the 29 carbon atoms can be recognized.

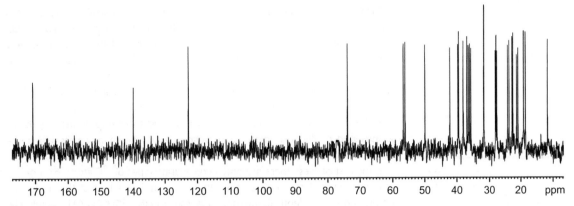

Figure 4.24 ^{13}C NMR spectrum (fully decoupled) of cholesterol acetate.

In the case of ^{13}C NMR spectra, unless special precautions are taken the peak heights are not strictly proportional to the relative numbers of different carbons in the molecule and therefore the integration is not routinely presented. However, the observed peak heights often serve as a useful guide to the relative abundances.

^{13}C chemical shifts

^{13}C chemical shifts, like ^{1}H NMR shifts, are also found to be diagnostic of structure, and typical ranges of ^{13}C shifts for different types of compound and functional groups are given in Figure 4.25. The shifts are governed by the same factors discussed earlier for protons.

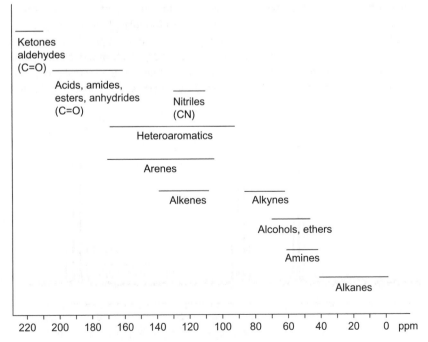

Figure 4.25 Typical ^{13}C NMR chemical shift ranges.

Carbonyl groups, carbons in aromatic rings, alkene, and alkyl fragments can usually be readily distinguished. For example, in the spectrum shown in Figure 4.23, the absorption at δ 210 is typical of the carbonyl group in the ketone. Likewise, the C=C carbon atoms in cholesterol acetate are characterized by the peaks at δ 140 and δ 120 in Figure 4.24. And the high resolution achieved for decoupled ^{13}C spectra shows that for arenes, for example, all the aromatic carbon atoms can often be distinguished (whereas for aromatic *protons* there is usually considerable overlap).

At this stage you are encouraged to try a self-test to attempt to identify the compounds whose fully decoupled ^{13}C NMR spectra are shown in Figures 4.26 and 4.27 and for which the appropriate formulae are provided.

In the first of these there are only two separate types of ^{13}C atom (with one peak approximately twice as high as the other). This indicates that of the three carbon atoms, two are in identical environments: the structure $ICH_2CH_2CH_2I$ would clearly fit, whereas the alternatives $CH_3CH_2CHI_2$ or CH_3CHICH_2I would each be expected to have three separate resonances.

In the second molecule, which clearly contains three carbon atoms in different environments, the peak at δ 172 is typical of a carbonyl group in, for example, an ester or acid: on the basis of this spectrum, and the formula, the structure $CH_3C(O)OCH_3$ is suggested.

Carbon–proton multiplicity determination

The main disadvantage of the otherwise helpful *full* decoupling procedure is that vital structural information which would result from the observation of C–H splittings is lost. Fortunately, it is also possible to record a ^{13}C NMR spectrum employing a partial decoupling procedure such that the splitting pattern between each carbon and the attached protons is effectively retained but further splittings are removed. This is called 'off-resonance' decoupling. The number of lines in the patterns and their relative heights are governed by the same simple rules formulated for $^{1}H-^{1}H$ splittings, and Pascal's triangle again provides the key guidelines. For example, the spectrum of butanone recorded in this way is shown in Figure 4.28, which should be compared with the fully decoupled spectrum in Figure 4.23. Note that the methyl-group carbon resonances at δ 28 and δ 7 are now split into quartets which are approximately

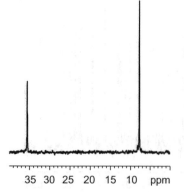

Figure 4.26 ^{13}C NMR spectrum (fully decoupled) of an unknown compound $C_3H_6I_2$.

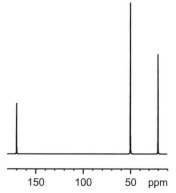

Figure 4.27 ^{13}C NMR spectrum (fully decoupled) of an unknown compound $C_3H_6O_2$.

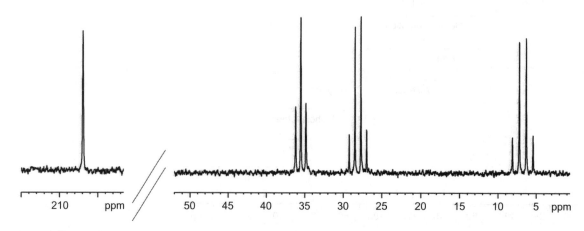

Figure 4.28 ^{13}C NMR spectrum (off-resonance decoupled) of butanone.

1:3:3:1 patterns, indicative of three attached protons in each group; the methylene carbon resonance at δ 36 becomes essentially a triplet (1:2:1) pattern whereas the resonance at δ 207, from C=O, has no splitting since this carbon has no attached protons.

This approach has now been largely replaced by a series of high sensitivity experiments which enable carbon–proton multiplicity to be assigned by inspection, the so-called 'spectral editing' method. Here, the complex series of radiofrequency pulses (referred to as a Distortionless Enhancement of Polarization Transfer, or DEPT, sequence) applied to the sample results in the phase of the detected signals being controlled by the carbon–proton multiplicity. The term phase is used to refer to whether the peak appears as a positive-going signal, above the base line, or a negative-going signal, below the base line. Normally, four proton-decoupled spectra are recorded, a normal ^{13}C spectrum, a DEPT-45° (i), a DEPT-90° (ii), and a DEPT-135° (iii). The first, normal spectrum contains signals from methyl (CH$_3$), methylene (CH$_2$), methine (CH), and quaternary (C) carbons. The remaining spectra only contain signals from carbons with protons directly attached. In case (i) all the carbon signals arising from groups with protons attached appear as positive-going signals, in case (ii) the only signals that are visible arise from methine carbons, while in case (iii) the methyl and methine (CH) signals appear as positive peaks while the methylene signals are negative. Using these four spectra collectively allows unambiguous multiplicity assignment. Figure 4.29 illustrates the DEPT spectra of butanone, for you to check the assignment of structure.

In summary, ^{13}C NMR spectroscopy with a pulsed (FT) spectrometer will provide a very valuable amount of information about an unknown sample: the number of different carbon atoms, their relative chemical environments, and also the number of hydrogen atoms attached to each carbon. Related experiments, with the same sample in the spectrometer but with a different frequency, will allow another nucleus (e.g. ^{1}H) to then be studied. So this enables information from, for example, both ^{1}H and ^{13}C NMR spectra for a given compound to be combined to provide crucial information about its molecular structure; examples for practice are provided at the end of the chapter.

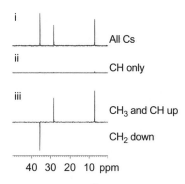

Figure 4.29 DEPT ^{13}C NMR spectra (fully decoupled) of butanone: the carbonyl carbon has δ 210.

4.8 2-Dimensional NMR methods

The development of NMR spectroscopy has led to its establishment as perhaps the most important and powerful method available for structure determination and is continuing in many exciting ways. The vast amount of detailed information to be obtained from ^{13}C NMR (from shifts, coupled and decoupled spectra) has led to the harnessing of developments in electronics and computing in the manufacture of pulsed spectrometers of enormous sophistication; in many of these the operator can now choose resonance frequencies from any of a wide range of nuclei with magnetic moments (^{1}H, ^{17}O, ^{19}F, ^{31}P, etc, as described on page 88, section 4.6). Other developments have facilitated the linking of these resonances as 2-dimensional presentations (2-dimensional NMR) in order to aid the characterization of complex molecules. Whilst the technical and theoretical details of these approaches are beyond the scope of this book, it is hoped that a brief coverage will illustrate their scope and usefulness (as well as to stimulate further reading and investigation).

In the 2D experiments a combination of pulses is typically used. After the first pulse, the now-excited nuclei start to decay back to their starting state (say ^{1}H in a proton experiment) and it is this procedure that would generate the (usual) signal for

FT analysis and presentation. If instead, a short period of time later, a second pulse is applied (this could be to either ^{1}H or a second nucleus, e.g. ^{13}C) then we can begin to probe magnetic interactions between the excited nuclei because they evolve during this time period. The new information is collected, as in an ordinary FT experiment (although there will be differences depending on the ways in which we wish to view the nuclei interacting with each other—e.g. via spin-spin coupling between their magnetic moments, as described earlier). The main difference is that this experiment is repeated many times, with a slightly longer delay each time between the pulses, to produce a series of NMR spectra. When these are analysed together they can be plotted in 2-dimensions i.e. a ^{1}H NMR versus ^{1}H NMR spectrum or a ^{1}H NMR versus a ^{13}C spectrum. Examples of these types of spectra are shown in Figures 4.30 and 4.31.

In Figure 4.30 a ^{1}H–^{1}H 2D NMR spectrum of styrene is shown. The peaks that sit on the diagonal correspond to the normal ^{1}H spectrum (Figure 4.15), with all the chemical shift and splittings you would expect (these also appear as 1D spectra on the horizontal and vertical axes). We see *off-diagonal* peaks in the 2D spectrum when the nuclei interact with one another magnetically. Precise analysis of this information using the intercept positions on the axes indicates with which other atom(s) the nuclei are interacting. In this case, the stronger off-diagonal connections between H_A and H_B which have the larger HH coupling, are boxed to exemplify the effect. The presence of these cross-peaks confirms that protons H_A and H_B in styrene are connected by a mutual spin–spin

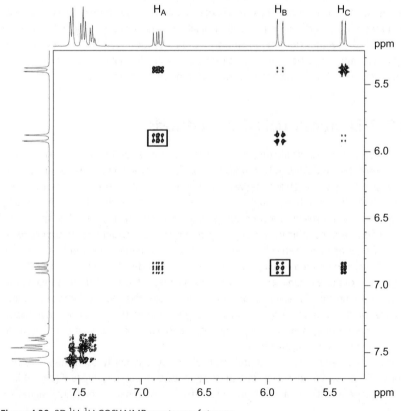

Figure 4.30 2D ^{1}H-^{1}H COSY NMR spectrum of styrene.

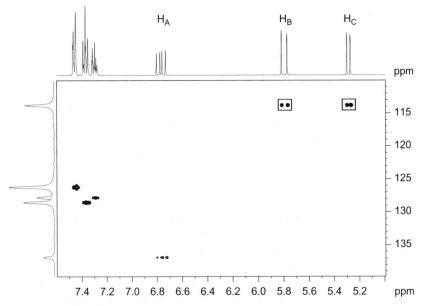

Figure 4.31 2D ^{1}H-^{13}C heteronuclear correlation NMR spectrum of styrene.

coupling, J_{HH}, which in this case is 17.6 Hz (usually measured from the 1D spectrum). This approach provides a particularly good way to check the 'connectivity' in a molecule, i.e. to establish which hydrogens are adjacent to which others. It is referred to as correlation spectroscopy and Figure 4.30 is an example of a so-called COSY spectrum.

For the heteronuclear 2D experiment shown in Figure 4.31, in which ^{1}H and ^{13}C spectra of styrene (phenylethene) are shown and plotted against each other, the peaks indicated allow you to work out which hydrogen is attached to which particular carbon atom, which can greatly assist in the use of NMR to obtain full details and confirmation of the structure of an unknown material. In this case, it is clear that hydrogen's H_B and H_C connect to a single ^{13}C resonance (with δ_C114) and are therefore attached to the same carbon centre (these resonances are boxed); proton H_A is attached to the carbon with δ_C137. You can also pick out the ^{13}C resonances from the ring carbons with protons attached (δ_C 126–128); the quaternary carbon (with no hydrogens attached) is not detectable in this experiment.

It is also possible to design further experiments using such pulse techniques to reveal interactions between atoms close to each other in space (the so-called Nuclear Overhauser Effect).

The combination of ^{1}H and ^{13}C NMR spectroscopy, with substantial further advances in resolution (e.g. in spectrometers with frequencies of up to 700 MHz and beyond and even NMR experiments with ^{15}N-labelled samples) provides a method for greatly enhanced potential for structure determination in very complex molecules, including proteins and enzymes. The technique allows researchers to study the detailed structure and conformation in solution (to complement solid-state studies by X-rays, as described in Chapter 6). The principles are illustrated by the NMR study of the peptide hormone angiotensin II, (whose structure is referred to on page 125, section 5.6; see for example, N. Zhou, G. J. Moore, and H. J. Vogel, *J. Protein Chem*, 1991, **10**, 333 and also G. A. Spyroulias, P. Nikolakopoulou, A. Tzakos, I. P. Gerothanassis, V. Magafa, E. Manessi-Zoupa, and P. Cordopatis, *Eur J. Biochem*, 2003, **270**, 2163 (see also the Online Resource Centre)).

4.9 Magnetic resonance imaging (MRI)

The development with perhaps the greatest implications of all is 'magnetic resonance imaging' or 'whole-body' NMR. This involves placing a much larger sample (e.g. a human body or limb) in the magnetic field and using the appropriate radiofrequency to obtain a resonance signal from, for example, ^1H or ^{31}P nuclei in the sample (for example, from water molecules in human cells or from phosphate in muscle enzymes). Signals from healthy and diseased tissue can often be distinguished and, as with X-rays, a map or image can be obtained.

In a simple ^1H NMR experiment we might have expected a single resonance from the H_2O which makes up the human body. However, for a three-dimensional object such as the human body we would expect the concentration of water to vary with tissue type and its spatial location. In magnetic resonance imaging this variation can indeed be explored, in experiments in which the irradiation frequency is held constant, and a series of measurements are carried out to encode the resulting signal intensity into a 3D map of position. Clearly, at any one particular position in the sample, the field is appropriate for ^1H resonance, and a signal 'reports' on the concentration of protons at that point. A series of experiments then reveals the variation of proton density across the sample and the information (image) is portrayed as a 3-dimensional voxel-based map of the signal across 3-dimensions. As with X-rays, a map or image can be obtained in which soft and hard tissues can be distinguished, as shown for example in Figure 4.32, in which the bones of a patient's spine are clearly seen. These studies, compared with the related application of X-rays, have the important advantage that the radiation employed is harmless.

Clinicians and spectroscopists working together have refined and extended these approaches to provide enhanced sensitivity and data handling techniques, some of which involve paramagnetic contrast agents. Advances in medical usage include the ability to monitor flow rates and oxygenation levels of specific organs and also the recording of high-resolution spectra (MRS, Magnetic Resonance Spectroscopy) at specific points of the body to follow the concentrations of specific biochemicals, such as

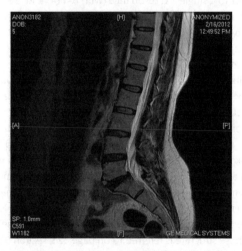

Figure 4.32 MRI scan showing the bones in a patient's spine.

pyruvate, lactate and choline (the 2-hydroxy-ethyltrimethylammonium ion, an essential nutrient and precursor to the neurotransmitter acetylcholine). Examples which describe the potential of its use as a non-invasive diagnostic test include the mapping out of spatial distribution of the concentrations of metabolites within the brain, for example to explore the possible association between tumour grade (and type) and detected levels of choline; and the monitoring of pyruvate concentrations in investigations of prostate cancer.

4.10 Summary

As a result of studying the material presented in this chapter you should have developed an understanding of the basic principles which underpin NMR (nuclear magnetic resonance) spectroscopy and its application to structure determination.

These key features include:

- the recognition of the phenomenon of nuclear spin (and associated nuclear magnetic moments) and identification of those nuclei (and isotopes) that possess these;

- the principles of nuclear magnetic resonance, with essential features of the instrumentation used to generate spectra;

- the appreciation of the importance of *chemical* shifts (and the influence of electronic effects) and *spin–spin splittings* in producing high-resolution NMR spectra which reflect the details of molecular structure (especially in relation to ^{1}H and ^{13}C nuclei);

- more detailed application to the study of, for example, processes involving rapid interconversion and the production of detailed 2-dimensional spectra which are particularly rich in structural information.

You should also have developed basic problem-solving skills in the use of NMR spectroscopy (which can be used together with other spectroscopic techniques, like IR and UV, and mass spectrometry) for solving structural problems.

You should also have developed an initial appreciation of the use of NMR spectroscopy in a wide range of applications in chemistry and biology, especially its use in Magnetic Resonance Imaging (MRI) and its potential for extremely wide applications in medical scanning.

4.11 Exercises

4.1 Identify the compound, of molecular formula C_3H_8O, whose ^{1}H (a) and ^{13}C (b) NMR spectra are shown in Figure 4.33. DEPT experiments provide information on the number of protons attached to the carbon; the results are indicated in brackets.

4.2 Identify the compound, of molecular formula $C_4H_8O_2$, whose ^{1}H (a) and ^{13}C (b) NMR spectra are shown in Figure 4.34. DEPT experiments provide information on the number of protons attached to the carbon atoms; the results are indicated in brackets.

4.3 In the NMR spectrum of butanone (Figure 4.11) recorded on a 400 MHz spectrometer, determine the separation between the 1:2:1 peaks of the methyl

Worked solutions to the exercises are available on the Online Resource

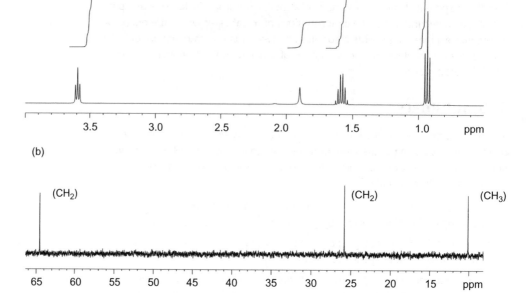

Figure 4.33 ^{1}H NMR and ^{13}C NMR spectra of an unknown compound, C_3H_8O: Exercise 4.1.

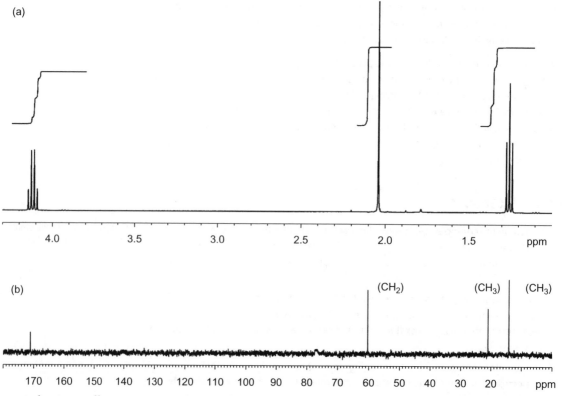

Figure 4.34 ^{1}H NMR and ^{13}C NMR spectra of an unknown compound, $C_4H_8O_2$: Exercise 4.2.

group CH$_3$ resonance at δ 1.05 ppm (in δ units). Hence calculate the coupling Δν, (in Hz) between the methylene and methyl hydrogen atoms in the ethyl group in the molecule.

4.4 The ^{1}H NMR spectrum of the compound whose IR spectrum is shown in Exercise 2.13, page 102 and whose mass spectrum is given in Exercise 5.3 (page 127) consists of a multiplet at δ 7.2 (area 5), a singlet at δ 5.1 (area 2), and a singlet at δ 1.95 (area 3). What can you conclude from this information about the structure of the molecule? When you have studied the mass spectrum and IR spectrum too, can you suggest an overall structure? What features would you expect to find in the ^{13}C NMR spectrum to enable you to add further evidence to substantiate your conclusion?

4.5 Figure 4.17 shows the ^{1}H NMR spectrum of cumene (2-propylbenzene). The complex pattern at δ 7.2, with some distortion, is from the protons of the phenyl ring. Using the expanded spectrum (inset) for the peaks between 7.1 and 7.3 ppm, analyse the pattern of shifts, splittings, and integrations in order to assign shifts and splittings to individual protons in the ring.

4.6 The ^{1}H NMR spectrum of cyclohexane recorded on a 500 MHz spectrometer comprises a single line at room temperature at δ 1.44. As the temperature is lowered, the spectrum changes to give, at −80 °C, a 1:1 pattern where the separation between the lines is *ca.* δ 0.48. Explain why two absorptions are seen and provide an explanation as to why coalescence occurs as the temperature is raised.

4.7 Predict the appearance (splitting patterns) of the NMR spectra (both ^{31}P and ^{19}F) for the molecule PF$_3$, which is known to be pyramidal.

Note: Further problems which incorporate information from integrated spectro-scopic experiments, (Mass spectrometry, IR and UV spectroscopy, ^{1}H and ^{13}C NMR spectroscopy) can be found at the end of the next chapter (page 128).

4.12 Further reading

L. M. Harwood and T. D. W. Claridge (1996), *Introduction to Organic Spectroscopy*, Oxford University Press, Oxford.

Peter Hore, Jonathan Jones, and Stephen Wimperis (2015) , *NMR: The Toolkit*, 2nd Edition, Oxford University Press, Oxford.

D. H. Williams and I. Fleming (2007) , *Spectroscopic Methods*, 6th Edition, McGraw-Hill, Maidenhead.

5 Mass spectrometry

5.1 Introduction

The technique of mass spectrometry, which owes its origin to pioneering experiments carried out at the beginning of the twentieth century, is now established as a remarkably versatile and effective method for obtaining the relative molar masses, formulae, and structures of molecules. As we will show in this chapter, its application enables many sophisticated structural problems to be solved rapidly, even when only minute quantities of material are available. And it is a particularly powerful technique when used in structure determination in combination with NMR spectroscopy and other techniques, and finds increasing use in environmental and biochemical applications.

This chapter sets out firstly to introduce you to the fundamental principles of mass spectrometry (which differ substantially from those encountered in the spectroscopic techniques described in the previous chapters). We then aim to explain and exemplify the use of this technique in the determination of molecular masses and formulae, as well as the interpretation of fragmentation patterns as a means to the determination of structure. We will provide some worked examples of structural analysis, to show you how problems can be solved using this technique and approach, and also provide some spectra from unknown compounds for you to tackle. Emphasis will also be placed on the use of this technique with those described earlier in the book (e.g. NMR and infra-red spectroscopy), which provide complementary information about the molecules under investigation. In the last section, we will survey the very wide range of structural and analytical applications of mass spectrometry across the chemical and biochemical sciences, with emphasis on its extreme sensitivity.

5.2 The mass spectrometry experiment

The principle of the method is to obtain a positively charged ion characteristic of the substance under investigation, and then to determine the *mass* of this ion using an approach closely related to that employed by J. J. Thomson for measuring the charge-to-mass ratio *(z/m)* for electrons. The procedure involves the use of electric and magnetic fields to deflect the charged particles.

Thomson and his colleague Francis Aston, both of whom were awarded Nobel Prizes, used a magnetic field to deflect a beam of positive ions, obtained by the ionization (loss of an electron) of neon atoms. Close examination of the trace produced by the positive ions as they impinged on a detector demonstrated that there were

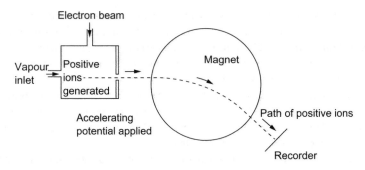

Figure 5.1 Basic features of a simple mass spectrometer, showing the focusing of positive ions of a given mass (*m*) and charge (*z*) on to the detector.

two different types of ion, characterized as those from the two neon **isotopes** (^{20}Ne and ^{22}Ne), which differ in mass because of the different numbers of neutrons in their nuclei.

The design of a simple mass spectrometer is shown in Figure 5.1. A very small amount of the vapour of the substance to be studied (obtained, for example by heating the sample) is introduced into the ionization chamber at very low pressure (about 10^{-4} N m^{-2}). The vapour is bombarded with high-energy electrons, and the collision between an electron and the molecule (or atom), M, causes an electron to be ejected, leaving a positively charged ion (M$^+$) in a process referred to as *Electron Impact (EI)*. The ions thus created in the ion source are attracted by an applied electrostatic potential and are hence accelerated towards the negative plate. The ions pass through a slit in the plate and into a magnetic field: the positive ions then become deflected by an amount which depends upon their *mass (m)* and the *charge (z)*. The lighter the ion and the greater its charge, the greater will be the deflection. This type of mass spectrometer is often referred to as a (magnetic) sector instrument.

The derivation of the exact relationship is as follows: for an acceleration potential *V*, the *potential energy* of an ion, of charge *z*, generated in the ion-chamber is *zV*. The ion is accelerated through the slit, and in this process its potential energy, *zV*, is completely converted into *kinetic energy, mv²/2*, where *m* and *v*, are the mass and velocity, respectively, of the ion (Equation (5.1)).

When the positive ion passes into the magnetic field (of magnetic flux density *B*) it experiences a force at right angles both to the direction of motion and to the field direction. The magnitude of this force is *Bzv*. The positive ion is now constrained to move in the arc of a circle of radius *R*, as given in Equation (5.2).

Combination of these equations leads to the important expression (Equation (5.3)) linking *m*, *z*, *B*, *R*, and *V*. This equation shows that for an ion of given mass (*m*) and charge (*z*), the radius of the circle of motion (*R*) is determined by *B* and *V*, i.e. the magnitudes of the magnetic and electric fields. In practice, *R* remains fixed by the geometry of the apparatus and the position of the detector. It can then be seen that if *V* is kept constant and *B* is varied, Equation (5.3) will be satisfied for ions of different *m/z* for different values of *B*. The value of *B* needed to get a particular type of ion to be deflected to the recorder is a measure of *m/z* for that ion.

Most positive ions generated will have lost just one electron and they will therefore have the same charge (opposite in sign, but equal in magnitude, to that of the

Ionization

$$M \rightarrow M^+ + e^-$$

$$zV = \frac{mv^2}{2} \tag{5.1}$$

$$Bzv = \frac{mv^2}{R} \tag{5.2}$$

$$\frac{m}{z} = \frac{B^2 R^2}{2V} \tag{5.3}$$

electron). This means that in an experiment, as *B* is varied, ions of different *mass* arrive at the recorder (at different values of *B*) and a read-out of the masses of the various ions concerned can be plotted. Since the mass of the electron is very small compared with the mass of the nucleus, the experiment is effectively determining the masses of the parent molecules. An electron-multiplier usually serves as a collector and detector of positive ions, and the arrival of the ions gives rise to a signal. This procedure produces a **mass spectrum**—essentially a plot of the masses of the positive particles present against the relative number of ions of each mass. The scan is calibrated with a peak from a substance of known relative molar mass.

Early mass spectrometers like this achieved a low resolution (e.g. separation of two fairly similar masses, such as ^{20}Ne and ^{22}Ne). More modern mass spectrometers have been designed with an extra focusing system based on an *electrostatic* or *quadrupole analyser*; substantially enhanced resolution can then be achieved, by obtaining ion beams with more precisely defined energies. An alternative type of *mass analyser*, now very widely employed, involves the use of the *Time-of-Flight method* (*TOF*): once the ions have been generated and accelerated (as described above), they move across a field-free region at different velocities defined by Equation (5.4) (derived from Equation (5.1)). They will arrive at the detector (see Figure 5.2) at different times after generation—and the times will be dependent upon their masses. Their separation, after measurement of *t*, can lead to a read-out of different masses.

$$v = \sqrt{\frac{2zV}{m}} \qquad\qquad (5.4)$$

Mass spectrometers have also been designed with alternative *ion sources*, as well as *mass analysers*, and in a number of these the positively charged species detected is the protonated molecule (MH$^+$) produced from the parent molecule (M). These approaches to the generation of the positive ions include the use of *Fast Atom Bombardment* (*FAB*) to volatilize and ionize involatile inorganic solids (e.g. with a beam of caesium ions), and in *MALDI* spectrometers the molecule under investigation is

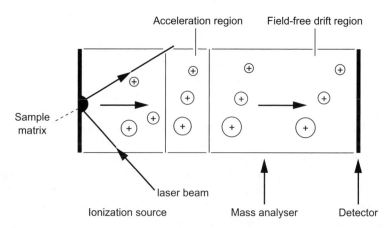

Figure 5.2 Simple diagram of a Time-of-Flight mass spectrometer, with MALDI method for positive ion generation.

embedded in a polymer matrix and subsequently vaporized by a blast from a laser (*Matrix Assisted Laser Desorption Ionization*) before proton transfer (see Figure 5.2). In an *Atmospheric Pressure Ionization Spectrometer* (*API*), a solution containing the substance is heated rapidly in a corona discharge; in *chemical ionization* (*CI*) spectrometers an inert gaseous substrate (e.g. CH_4) is first excited with subsequent proton transfer to the substrate. In an *Electrospray Ionization* (*ESI*) *Mass Spectrometer*, an aqueous solution is evaporated leaving droplets of molecules in the inlet chamber; protonation (e.g. of a biological sample) gives positive ions ready to be focused and detected.

The examples described later demonstrate the remarkable power of modern mass spectrometers to attain extremely high resolution (separation of closely similar masses), and determination of molar masses to great precision, often achieved very rapidly and with great sensitivity. For example, spectra can now be obtained from amounts of material in the 10^{-12}–10^{-15} g range; and measurements of accurate mass can routinely be made down to the third or fourth place of decimals.

5.3 Measuring relative molecular and atomic masses

Modern mass spectrometers can be used in various kinds of study. For example, they may be employed to give fairly rapid scans of the relative masses of the ions from a variety of substances, for example as obtained in *Electron Impact (EI)* experiments. Alternatively, under conditions of high resolution, they can be used to separate closely spaced peaks and to determine the appropriate relative atomic and molar masses with precision. The following example illustrates the advantages of the latter approach if the maximum amount of information is to be derived. Thus, a peak which corresponds to mass 28 might be due to nitrogen ($^{14}N_2$), carbon monoxide ($^{12}C^{16}O$), or ethene ($^{12}C_2{}^1H_4$). However, these three molecules have slightly different relative molecular masses, as shown in the marginal note (these are based on the internationally accepted scale, with 12 exactly for the ^{12}C isotope). At this stage you are encouraged to use this information to work out the atomic weights of ^1H, ^{12}C, ^{14}N, and ^{16}O to four places of decimals, for later use.

A high-resolution mass spectrometer can readily be used to identify a peak exactly enough for it to be characterized as the positive ion from one of these. Furthermore, if all three substances were to be present together, then under high-resolution conditions three separate peaks could be resolved. If one relative molar mass is accurately known, this can be used to calibrate the field scan so that the other molecular masses can also be accurately determined.

Relative heights of separate peaks can also be used to obtain quantitative information. For example, from the mass spectrum of neon can be measured not only the relative atomic masses of the constituent isotopes (^{20}Ne, ^{21}Ne, ^{22}Ne), to an accuracy of approaching 1 part in 10^6, but also the relative abundance of the separate isotopes in the mixture.

We should also remember to distinguish the *separate* isotopic atomic masses measured with the mass spectrometer from the *weighted average* obtained by other (chemical) methods. For example, ^{35}Cl has a relative atomic mass of 34.9688, and that of ^{37}Cl is 36.9659; the average atomic mass of the natural mixture of isotopes (75.53% ^{35}Cl, 24.47% ^{37}Cl) is 35.45.

Relative molar masses	
$^{14}N_2$	28.0062
$^{12}C^{16}O$	27.9949
$^{12}C_2{}^1H_4$	28.0313

Relative atomic mass	Relative abundance (%)
^{20}Ne 19.9924	90.92
^{21}Ne 20.9940	0.26
^{22}Ne 21.9914	8.82

5.4 Mass spectrometry of molecules: a detailed example

When an organic compound is introduced into an *EI* mass spectrometer, the molecules become ionized, by the loss of an electron, and the positive ions produced pass through the focusing system, leading usually to a peak at the appropriate relative molar mass. However, the mass spectrum of an organic compound also contains extra information about fragmentation which can be extremely useful.

For example, Figure 5.3 shows the mass spectrum of ethanol (CH_3CH_2OH): this is a plot of signal height (proportional to the number of ions of given *m/z*) against increasing *m/z* (since nearly all the ions have the same unit charge this axis effectively corresponds to increasing mass).

Most peaks occur at (or close to) integral values (the extra precision possible with high resolution is not usually employed at this stage). Many of the peaks are derived from ethanol by processes which will be described shortly.

There may also be peaks due to traces of air in the instrument: this can give rise to signals from N_2 *(m/z* 28) and O_2 *(m/z* 32), approximately in the expected ratio 4:1. These peaks may be used to calibrate the scan. The peak heights are expressed as percentages of the height of the highest peak (called the **base peak)**, which in this example is the peak with *m/z* 31.

The spectrum shows the expected peak at *m/z* 46, corresponding to the **molecular ion** (M^+) of the parent molecule (the relative molecular mass is the sum of the relative atomic masses of $2C + 6H + O$). Since this has lost an electron compared to the original molecule, it is usually referred to with a positive charge and a dot (single electron) as follows $[C_2H_5OH]^{+\cdot}$.

There is also a very small peak at *m/z* 47, called the (M + 1) peak, which corresponds to the relatively few ethanol molecules present which, because they contain a ^{13}C, ^{17}O, or ^2H atom, have a molar mass of 47. ^{13}C has a natural abundance compared with ^{12}C of 1.1%; for ^{17}O, relative to ^{16}O, the figure is 0.04% and for ^2H, relative to ^1H, the abundance is 0.01%. An even smaller (M + 2) peak arises from the molecules which contain two ^{13}C atoms, or an ^{18}O atom, or a ^{13}C atom and an ^{17}O atom, etc.

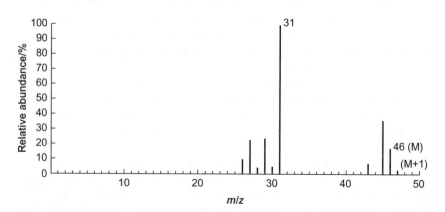

Figure 5.3 Mass spectrum of ethanol.

Other peaks in the spectrum occur because some ethanol molecules which are ionized to give M⁺ then break down (*fragment*) to give smaller positive ions, a process which is understandable in terms of the high energy of the bombarding electrons. The positive fragments are also accelerated and focused to be collected and recorded for their particular values of *m/z*. The large peak in the mass spectrum of ethanol (the base peak) is at *m/z* 31: this corresponds to the fragment $[CH_2OH]^+$ obtained by loss of •CH_3 from the parent ion $[CH_3CH_2OH]^+$. The structure of positive ions will be discussed later, as will some guidelines for interpreting fragmentation patterns, but it should at this stage be apparent that these peaks contain important clues about the molecular structure.

It should be noted at this stage that other methods of producing positive ions for study (e.g. the *ESI, MALDI, CI,* and *APCI* techniques) often produce spectra with a smaller degree of fragmentation than induced by *EI* techniques.

5.5 Analysis of mass spectra

The molecular ion: determination of relative molar mass and formulae

Many molecules give a peak of appreciable size for the molecular ion. It is usually a reliable guide that a molecule with π-electrons (e.g. benzene) will give a detectable molecular ion (M) since one of these electrons can normally be lost (to give M⁺) without the breakdown of the bonding framework in the molecule. However, because in some cases no peak from a molecular ion can be observed, care must be taken before it can be assumed that the peak at highest *m/z* is from the molecular ion.

It may be possible at this stage to determine very accurately the *relative molecular mass* of any given peak (if the high-resolution facility is available) and this will then be carried out for the molecular ion itself. Because different atoms do not have exactly integral atomic masses and, in addition, because various combinations of similar mass are not identical (contrast for example C_2H_4 and N_2) the exact relative molecular mass (to 3 or 4 places of decimals) should characterize the molecular formula exactly. For instance, a molecular peak with *m/z* at almost exactly 60 could be from various possible molecules with different formulae, including $C_2H_4O_2$ (e.g. ethanoic acid (acetic acid, CH_3CO_2H)) and C_3H_8O (e.g. propanol). Under conditions of high resolution, these possibilities can be clearly distinguished (as shown in the margin). Indeed, given a precisely determined relative molecular mass we can obtain the molecular formula: for example, a molecular ion with *m/z* 94.0419 can be reliably attributed to a compound with the molecular formula C_6H_6O.

Formula	Relative molar mass
$C_2H_4O_2$	60.0211
C_3H_8O	60.0575

Information can also often be extracted from the M, (M+1), and (M+2) peaks even if a high resolution facility is not available. For example, for ethanoic acid ($C_2H_4O_2$), the height of the (M+1) peak at *m/z* 61, which is mainly due to $^{12}C^{13}CH_4O_2$, should be just over 2% of the height of the peak from the molecular ion (*m/z* 60). This is because there is approximately a 2.2% chance that a molecule will contain one ^{13}C atom (there will be a much smaller percentage of molecules containing ^2H or ^{17}O). The corresponding figure for C_3H_8O is just over 3%, and for a compound with, say, eleven carbon atoms the relative intensities of M:(M+1) peaks should be about 88:12 (i.e. $(100-(11\times1.1)):(11\times1.1)$). Clearly, then, measurement of the *relative heights* of the M and (M+1) peaks, and sometimes of the (M+2) peak can be diagnostically useful, and extensive tables of accurate M:(M+1):(M+2) ratios for various molecular formulae are available. In any case, from a brief inspection, sensible deductions can usually be made. Thus the (M+1) peak will be approximately $N\%$ of the main peak if the formula has N carbon atoms, a larger-than-usual M+2 peak may indicate that a sulfur atom is present in the molecule (^{34}S has a natural abundance of 4.22%), and so on.

In certain molecules the effect can be particularly striking. For example, Figure 5.4 shows the mass spectrum of chloromethane in which the peaks at m/z 50 and 52 are from the molecular ions of $CH_3{}^{35}Cl$ and $CH_3{}^{37}Cl$, respectively, their relative intensities being in the ratio expected from the relative isotopic abundances of ^{35}Cl and ^{37}Cl (approximately 3:1). For bromomethane (Figure 5.5) the two almost equally intense peaks are from $CH_3{}^{79}Br$ and $CH_3{}^{81}Br$, the two bromine isotopes having approximately the same abundance.

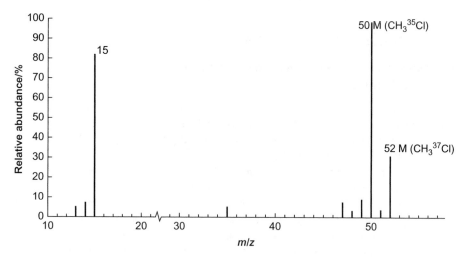

Figure 5.4 Mass spectrum of chloromethane, CH_3Cl.

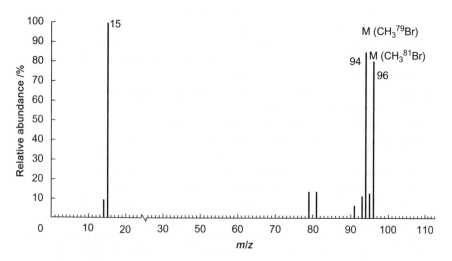

Figure 5.5 Mass spectrum of bromomethane, CH_3Br.

Another useful 'rule' is that a molecular ion with an *odd* value of *m/z* generally characterizes a molecule with an odd number of nitrogen atoms: you are recommended to check for yourself that this works (e.g. for compounds such as CH_3NH_2, CH_3CONH_2, $C_6H_5NH_2$, say).

It should also be remembered that a peak of reasonable intensity at the highest *m/z* value observed is not necessarily from the molecular ion, but may be instead from part of the fragmentation pattern of a compound whose molecular ion has a peak too small to be clearly established. Wherever possible, therefore, data should be interpreted together with information from other spectroscopic techniques and from conventional molecular mass and both empirical and molecular formulae determination.

Fragmentation patterns of organic molecules: diagnosis of structural features

A variety of fragmentation pathways is normally possible for M^+ and for each route, one of the fragments retains the positive charge. For example, another possibility here is fragmentation to $P+Q^+$ and further fragmentation of either P^+ or Q^+ may also occur.

Fragmentation

$$M \rightarrow M^+ \rightarrow P^+ + Q$$

The recognition of preferred modes of fragmentation (e.g. whether for this molecule a larger peak for P+ or Q^+ is obtained) is assisted by practice with spectra from known molecules, but deductions are based mainly on chemical intuition, and a few simple guidelines. It also helps that most of the principles involved in recognizing and predicting fragmentation patterns are closely related to those employed for discussing the chemistry of reactions in solution. For instance, we will need to consider which of a variety of possible fragments is best able to bear a positive charge, which bond is the weakest and therefore most likely to break, and which relatively stable entities might readily be formed in simple decomposition pathways.

First, remember to do some elementary book-keeping with electrons and charges; almost all molecules have an even number of (paired) electrons, so that the positive molecular ion formed by ionization must have not only a *charge* but also an odd number of electrons. The fate of both the charge and the unpaired electron should be considered when fragmentation patterns are being interpreted.

The main types of fragmentation possible for a molecule may be characterized as follows.

(i) **Simple cleavage.** This involves the breakage of a single bond in the molecular ion, and a good example is provided by the mass spectrum of ethanol (Figure 5.3). The molecular ion is at *m/z* 46 and the *base peak* is at *m/z* 31; the latter corresponds to a molecular ion which has lost a group of mass 15 before being accelerated and focused. It is described as an (M−15) peak (i.e. M minus 15), and is due to the ion $[CH_2OH]^+$ formed as shown.

$$[CH_3CH_2OH]^{+\bullet} \quad m/z \ 46$$

$$\downarrow$$

$$\cdot CH_3 \ + \ [CH_2OH]^+$$
$$m/z \ 31$$

Note that the 'dot' indicates an unpaired electron. The products of this fragmentation then are the charged *ion* $[CH_2OH]^+$ (since the charge resides effectively on carbon this type of species is sometimes called a carbonium ion or carbocation and written $^+CH_2OH$) and the neutral methyl radical $\cdot CH_3$. Only the former, being charged, is recorded in the mass spectrum. Some of the other peaks arise as shown. The peaks with *m/z* 27–29 are typical of a molecule containing an *ethyl* group, just as the appearance of an (M−15) peak is usually indicative of a *methyl* group in the parent compound.

$$[C_2H_5OH]^{+\bullet} \longrightarrow H\cdot \ + \ [C_2H_5O]^+$$
$$m/z \ 45$$

$$[C_2H_5OH]^{+\bullet} \longrightarrow HO\cdot \ + \ [C_2H_5]^+$$
$$m/z \ 29$$

$$[C_2H_5OH]^{+\bullet} \longrightarrow H_2O \ + \ [C_2H_4]^{+\bullet}$$
$$m/z \ 28$$

$$[C_2H_5]^+ \longrightarrow H_2 \ + \ [C_2H_3]^+$$
$$m/z \ 27$$

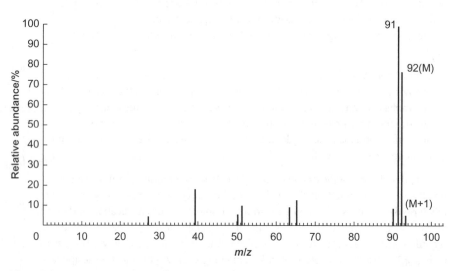

$^+CH_2-\overset{..}{O}-H \longleftrightarrow CH_2=\overset{+}{O}-H$

The reason that the peak with m/z 31, $[CH_2OH]^+$, is larger than those of the other positive ions is that this is a relatively *stable* positive ion compared to some of the other possible ions (e.g. $[CH_3]^+$, $[CH_3CH_2]^+$). This stability arises because oxygen has a lone-pair of electrons which can help stabilize the positive charge on the adjacent carbon atom: this is possible because there is a spreading (delocalization) of both the charge and the electrons between carbon and oxygen, a phenomenon which is indicated diagrammatically as shown. The use of the double-headed arrow implies that the actual electronic structure is somewhere in between the two extremes indicated.

A similar explanation accounts for the formation of the fairly intense peak with m/z 45, attributed to the ion $^+CH(CH_3)OH$ formed by loss of a hydrogen atom: this ion is more stable than $CH_3CH_2O^+$ which has the positive charge located solely on the electronegative oxygen. This mode of fragmentation is often observed when an ion can be produced with the positive charge on a carbon atom adjacent to an atom with a lone pair of electrons (e.g. O, S, N), and is therefore often significant for alcohols, ethers, thiols, and amines. Their typical fragmentation patterns often provide a means by which these molecules can be recognized.

Other cases where stabilized positive ions are produced include compounds containing the phenylmethyl (benzyl) group and also carbonyl-containing compounds. For example, Figure 5.6 is the mass spectrum from methylbenzene (toluene) which illustrates the behaviour of compounds in the former group. In addition to the molecular ion at m/z 92 and the (M+1) peak at m/z 93 (which has an intensity which is about 8% of the molecular peak, since there are seven carbon atoms in methylbenzene; see page 107, section 5.4), there is an intense peak (the base peak) at m/z 91 (M−1). This is from some of the methylbenzene molecules in the ionization chamber which lose first an electron and then a hydrogen atom, to give the phenylmethyl cation as shown.

The reason for the comparative stability of this cation is the ease with which the *aromatic ring* can delocalize the positive charge, a phenomenon which can be represented diagrammatically as shown.

Figure 5.6 Mass spectrum of methylbenzene, $C_6H_5CH_3$

This type of fragmentation, to give a peak with $m/z = 91$, is characteristic of compounds of the type $C_6H_5CH_2X$, for which this mode of cleavage is favoured. The mass spectrum from methylbenzene also shows other small peaks which indicate that alternative modes of fragmentation after high-energy bombardment are possible. These include the formation of two-, three-, and four-carbon fragments (that with m/z 51 is $[C_4H_3]^+$) from the aromatic ring; the occurrence of these, though helpful, is not as diagnostically useful as the evidence from the main pathway.

Carbonyl-containing compounds $RC(O)R'$ tend to decompose to give fragment ions in which the positive charge is again shared between carbon and oxygen (RCO^+). Other peaks arise because fragmentation occurs at the other C-alkyl bond to give $R'CO^+$ and because these ions readily lose carbon monoxide. For example, the mass spectrum from butanone (methyl ethyl ketone, Figure 5.7) shows the molecular ion (m/z 72) and the peaks at m/z 57 (M–15) and 43 (M–29), diagnostic of loss of $\cdot CH_3$ and $\cdot CH_2CH_3$ from the parent positive ion. The peak at m/z 29 ($^+CH_2CH_3$) evidently arises via loss of CO from $CH_3CH_2CO^+$. The peaks at 57 (M–15) and 29 (M–43) help characterize a *methyl-substituted* ketone (loss of $\cdot CH_3$ and both $\cdot CH_3$ and CO, respectively).

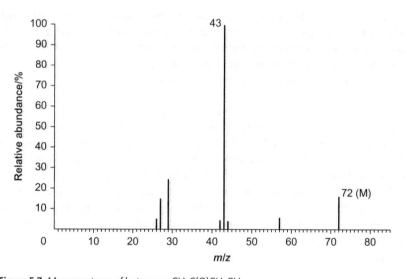

Figure 5.7 Mass spectrum of butanone, $CH_3C(O)CH_2CH_3$.

The spectrum of dimethylbutanone (methyl t-butyl ketone), Figure 5.8, shows the (M–43) peak quite clearly; the dimethylethyl (t-butyl) carbonium ion $^+C(CH_3)_3$, m/z 57, gives a particularly intense peak because a *tertiary* carbonium ion (with three alkyl groups attached to the carbon bearing the positive charge) is formed. This type of fragmentation is favoured here because a *tertiary* carbonium ion is more stable than a *secondary* carbonium ion, which is more stable than a *primary* carbonium ion (because of the overall electron-donating property of alkyl groups). This means that fragmentation is generally preferred at the point of branching.

(ii) **Fragmentation with rearrangement.** Occasionally, a fragmentation process is detected which is rather more complicated than those discussed in section (i) because molecular rearrangements are involved.

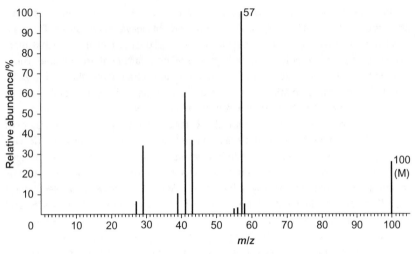

Figure 5.8 Mass spectrum of dimethylbutanone, $(CH_3)_3CC(O)CH_3$.

An example is provided by the mass spectrum of methyl butanoate ($CH_3CH_2CH_2CO_2CH_3$), which is shown in Figure 5.9. There is a trace of the expected molecular peak at m/z 102, and cleavage of the bonds to the carbonyl group leads to the peaks at m/z 71 (loss of $\cdot OCH_3$), and subsequently 43 (via loss of CO), and 59 (loss of $\cdot CH_2CH_2CH_3$). Other fragmentations lead to loss of a methyl group, to give the ion with m/z 87, and to $^+CH_2CH_3$ (m/z 29). However, the unusual peak is that at m/z 74 (M–28) which is thought to arise by transfer of a hydrogen atom to the suitably placed oxygen at the same time as fragmentation (the product ion (m/z 74) is recognizable as the *enol* form of methyl ethanoate, $CH_3CO_2CH_3$: see page 63, section 3.5). Note that peaks formed from single fragmentation patterns, as illustrated earlier, are usually *odd*: this *even* number loss (i.e. loss of a molecule rather than fragmentation of one bond) is characteristic of this behaviour.

This is known as a **McLafferty** rearrangement and tends to occur when a hydrogen atom and a carbonyl oxygen come into close proximity. Inspection of structural models reveals that this can be achieved with the minimum of strain when there are six atoms in the chain.

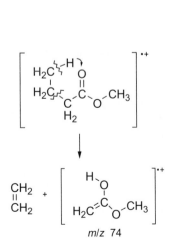

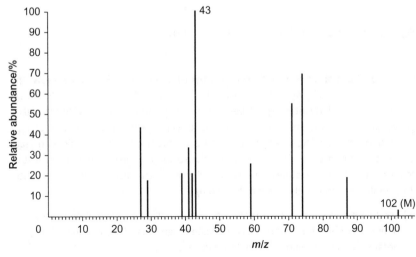

Figure 5.9 Mass spectrum of methyl butanoate, $CH_3CH_2CH_2C(O)OCH_3$.

Some examples of fragmentation patterns

The simple rules laid down so far should provide assistance with the solving of a variety of mass spectra. The following examples to a considerable extent typify the class of organic compound to which they belong.

Chloroethane, (Figure 5.10). The mass spectrum shows the two expected molecular ions from $CH_3CH_2{}^{35}Cl$ and $CH_3CH_2{}^{37}Cl$, at m/z 64 and 66, respectively (in the intensity ratio 3:1). The other main peaks are formed as shown. The loss of $\cdot H$ and $\cdot CH_3$ in the first two paths leaves the positive charge next to the chlorine (cf. $^+CH_2OH$, $^+CH(CH_3)OH$, page 110, section 5.5). Another mode of fragmentation in this example involves elimination of the neutral molecule hydrogen chloride, to give a peak at m/z 28 from the positive ion from ethene (ethylene).

$$[CH_3CH_2Cl]^{\cdot+}$$

- → $\cdot H$ + $^+CHClCH_3$ m/z 63, 65
- → $\cdot CH_3$ + $^+CH_2Cl$ m/z 49, 51
- → $\cdot Cl$ + $^+CH_2CH_3$ m/z 29
- → HCl + $[C_2H_4]^{\cdot+}$ m/z 28

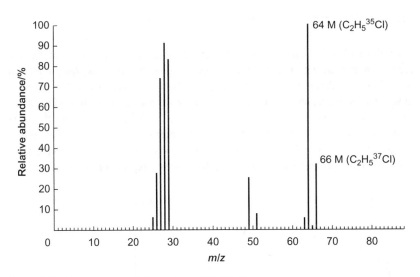

Figure 5.10 Mass spectrum of chloroethane, CH_3CH_2Cl.

Diethyl ether (ethoxyethane), $CH_3CH_2OCH_2CH_3$ (Figure 5.11). The spectrum shows a molecular ion at m/z 74, and an (M–15) peak at m/z 59 which indicates loss of a methyl group: this leaves the fragment $^+CH_2OCH_2CH_3$, in which the oxygen atom is again able to exert a stabilizing influence. The peaks at m/z 45 (M–29) and 29 itself are consistent with the occurrence of an ethyl group in the molecule; loss of this group gives either $^+CH_2CH_3$ and $\cdot OC_2H_5$, or $[C_2H_5O]^+$. The peak at m/z 31 is probably from $^+CH_2OH$, arising as shown. The (M–29) peak (m/z 45) probably arises via a similar process. The driving force for these fragmentations is the production of a stable molecule (ethene) with the retention of the positive charge next to oxygen in the remaining cation.

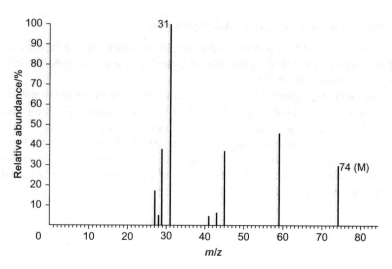

Figure 5.11 Mass spectrum of diethyl ether, $CH_3CH_2OCH_2CH_3$.

Diethylamine, $(C_2H_5)_2NH$ (Figure 5.12). The odd-numbered molecular ion (*m/z* 73) confirms a structure with a single nitrogen atom. Fragmentation in this example leads to peaks from $^+CH(CH_3)NHCH_2CH_3$ (*m/z* 72) and $^+CH_2NHCH_2CH_3$ (*m/z* 58 (M−15)), both of which have the positive charge adjacent to the nitrogen atom with its lone pair of electrons. Loss of an ethyl group is indicated by the (M−29) peak at *m/z* 44; this peak could be from $^+NHCH_2CH_3$, but the rearranged (stabilized) isomer $^+CH(CH_3)NH_2$ seems more likely, as with $^+CH(OH)CH_3$ in the previous example. The peak at *m/z* 30 is probably from $^+CH_2NH_2$, formed by a similar fragmentation–rearrangement process.

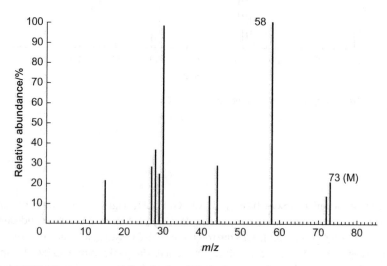

Figure 5.12 Mass spectrum of diethylamine, $(CH_3CH_2)_2NH$.

Acetophenone, $C_6H_5COCH_3$ (Figure 5.13). This mass spectrum shows a molecular ion at *m/z* 120 and a characteristic (M−15) peak (*m/z* 105) from loss of a methyl group to give $C_6H_5CO^+$. Fragmentation at the other side of the carbonyl group also occurs, to give the (M−43) peak at *m/z* 77.

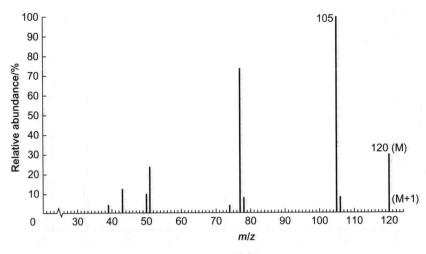

Figure 5.13 Mass spectrum of acetophenone, $C_6H_5C(O)CH_3$.

The detection of a peak at m/z 77 (and, in general, peaks in the 75–77 region) provides very strong evidence for a benzenoid compound. This is also true to a certain extent for the fairly prominent peaks at m/z 50 and 51 formed via the breakdown of the aromatic ring.

4-Methylpentan-2-one (methyl isobutyl ketone), $CH_3C(O)CH_2CH(CH_3)_2$ (Figure 5.14). In this example the peaks at m/z 85 (M–15), 43, and 57 (M–43) suggest the presence of the $COCH_3$ group. The peak from the $^+CH_2CH(CH_3)_2$ fragment (m/z 57) is not so dominant as that of $^+C(CH_3)_3$ in the isomeric ketone previously considered (Figure 5.8), which is as expected for the lower stability of the primary $[^+CH_2CH(CH_3)_2]$ rather than the tertiary $[^+C(CH_3)_3]$ ion. The peak at m/z 58 (M–42) arises via a McLafferty rearrangement, as shown (note this is an ion with *even* m/z; see earlier).

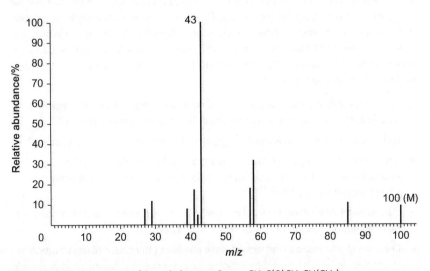

Figure 5.14 Mass spectrum of 4-methylpentan-2-one, $CH_3C(O)CH_2CH(CH_3)_2$.

4-Chlorobenzoic acid, $4\text{-}ClC_6H_4C(O)OH$ (Figure 5.15). This spectrum shows features as expected for a compound with two functional groups. The two peaks at high m/z values, 156 and 158, in the intensity ratio of approximately 3:1 are characteristic of

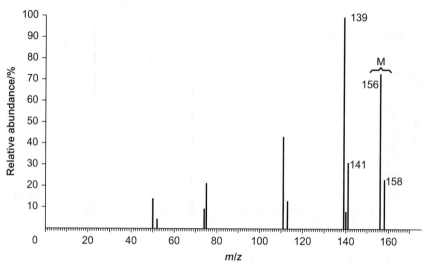

Figure 5.15 Mass spectrum of 4-chlorobenzoic acid, 4-ClC$_6$H$_4$C(O)OH.

a chlorine-containing compound. The peak at m/z 75 confirms the aromatic nature of the compound. The peaks at m/z 139 and 141, (M–17), again in the ratio 3:1, indicate that fragmentation has taken place (probably with loss of ·OH) with retention of chlorine in the positive ions (4-ClC$_6$H$_4$CO$^+$). The chlorine is also retained for the fragments at m/z 111/113 (4-ClC$_6$H$_4^+$, with both chlorine isotopes).

The mass spectra of the 2- and 3- isomers of this compound would resemble somewhat that of the 4-isomer and therefore mass spectrometry might not provide unambiguous assignment of a spectrum to one particular isomer. However, structure determination will usually be carried out with a variety of techniques and, for example, the nuclear magnetic resonance and infra-red spectra of the isomers, when examined in conjunction with the mass spectrometry evidence, would normally enable the distinction to be made.

At this stage, and before we attempt some worked examples for practice, let's just recap on the information which might help us solve the structure of an unknown molecule from its mass spectrum.

- Can we identify the molecular ion, and if possible, recognize the presence of, for example, chlorine or bromine (from their characteristic isotope pattern)?
- Is the molecular ion of *odd* mass (suggesting the presence of a nitrogen atom)?
- Can we work out, from the ratio of the M to M + 1 peak intensities, the approximate number of carbon atoms in the molecule (recall that the natural abundance of ^{13}C is 1.1% of ^{12}C)?
- Can we work out a likely molecular formula (if the experiment is carried out under high resolution conditions this can be revealed directly)?
- Does the C-to-H ratio in the formula give an idea of the extent of unsaturation in the molecule, which might then indicate the presence of aromatic rings or double bonds (e.g. as a starting point for further attempted analysis)?
- Can we recognize fragments with characteristic m/z values, e.g. 77 for a phenyl group, 29 could be an ethyl group, and noting that some peaks may be prominent because of resonance stabilization, e.g. benzyl C$_6$H$_5$CH$_2^+$?

- Does the loss of an *even* fragment suggest a rearrangement process (e.g. McLafferty rearrangement) or loss of CO (28) from RCO^+?

Taking all this into account, it will often be possible for you to suggest a molecular formula and appropriate structure, and test this by checking for the occurrence of major characteristic peaks from the anticipated fragmentation pattern.

Worked examples: determining the structure of organic molecules

Figures 5.16–5.18 are the mass spectra of three unknown compounds: the molecular ion (or ions) and the base peaks are denoted, with the ten most intense peaks in the fragmentation patterns. With no further information provided, can you identify the compounds? How might you attempt to confirm the formulae you have proposed? What other techniques might enable you to confirm the structures?

5.1 Figure 5.16 is the mass spectrum of ethylamine, $CH_3CH_2NH_2$. The important features to note are the odd-numbered molecular ion, indicating a molecule containing a single nitrogen atom, the intense (M–1) and (M–15) peaks, and the typical 'ethyl-group' peaks at 26-29. All this supports a molecular formula C_2H_7N (rather than say CH_3NO), which you would be able to confirm via the measurement of its molar mass to great precision at high resolution. The breakdown pattern observed derives, in part, from the ability of nitrogen to stabilize an adjacent carbonium ion, and loss of a stable molecule (ammonia in this example) is also observed. Note that the mass spectrum serves to distinguish clearly between this and the isomeric structure $NH(CH_3)_2$: the latter would not show significant ions at 26-29 nor, since $^+NHCH_3$ is not stabilized, at 30 (M–15). Ethylamine would also show characteristic 'ethyl group' peaks in the ^1H NMR spectrum, two different carbon atoms (via ^{13}C NMR spectroscopy) and an infra-red absorption from the N–H bond.

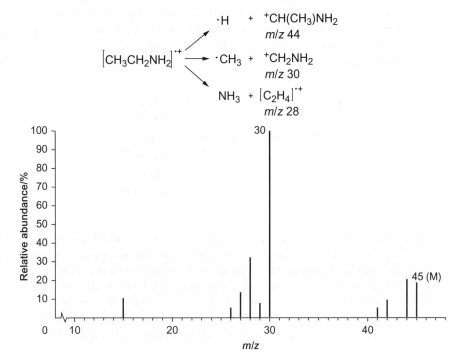

Figure 5.16 Mass spectrum of Worked example 5.1.

5.2 Figure 5.17 is the mass spectrum of a bromine containing compound. The crucial evidence here lies in the 'double' molecular peak, two units of m/z apart, from molecules containing ^{79}Br and ^{81}Br (in almost equal abundance). The peak at m/z 29 strongly suggests an ethyl group, which is confirmed by the peaks from m/z 25 to 28, leading to the structural assignment to bromoethane. Other fragments which can be recognized are the bromine ions themselves (m/z 79/81) and also the (M–15) peaks (m/z 93/95).

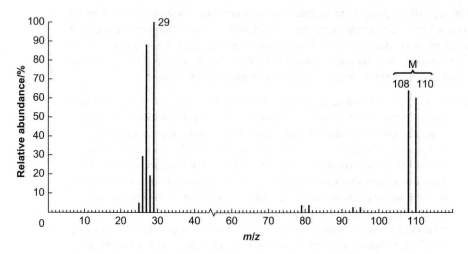

Figure 5.17 Mass spectrum of Worked example 5.2.

Comparison of this spectrum with that from chloroethane (Figure 5.10) indicates a greater relative extent of halogen loss in this example, which is consistent with the C–Br bond being weaker than the C–Cl bond. Again, the ^{1}H NMR spectrum would quickly confirm the presence of the ethyl group in the structure.

5.3 As judged by the mass spectrum in Figure 5.18, this compound is aromatic (there is a large peak at m/z 77, suggestive of C_6H_5, and also peaks at 50/51) and it can easily lose a single hydrogen atom (to give the large peak at 105; the relative molecular mass is 106). Two possibilities which could lose a fragment of mass 29 to leave the peak at m/z 77 are ethylbenzene and benzaldehyde. However, the former would be expected to give an intense peak at 91 (from phenylmethyl, $[C_6H_5-CH_2]^{+}$ see page 110, section 5.4 and

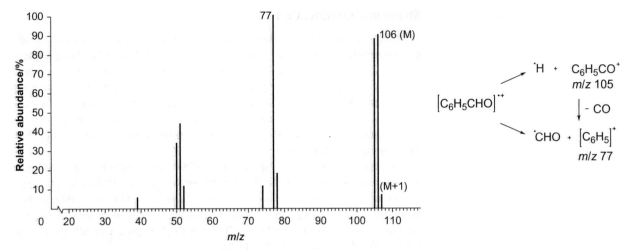

Figure 5.18 Mass spectrum of Worked example 5.3.

Figure 5.6), so the latter structure is preferred. The loss of a hydrogen atom is expected to be particularly facile for benzaldehyde.

To confirm the formula C_7H_6O, use high-resolution techniques to determine the accurate mass of the molecular ion and compare it with that calculated from accurate mass tables: as a self-test, why not check that this is different from that for $C_6H_5CH_2CH_3$? In practice, you would normally also have extra information available. In particular the ^1H NMR spectrum should confirm the 5 aromatic protons and the aldehydic proton and the ^{13}C NMR spectrum should confirm these features (see Table 4.2 and Figure 4.25): the carbonyl group stretching vibration should be very clear in the IR spectrum (see Table 2.5.).

5.6 Applications of mass spectrometry

A very wide range of mass spectrometers is now available commercially and these find very wide usage based on their remarkable ability to provide information on a molecular identification and structure determination as well as in analytical applications. Whilst the spectrometers typically operate with key principles and features in common (ionization source, mass analyser, and detector), developments have led both to large instruments for high-resolution work, with molecules of large molecular mass, and to miniaturized instruments capable of giving lower-resolution but rapid-response information on molecules with lower molecular weight. The latter instruments can sometimes be used as 'bench-top' spectrometers and used for sampling and analysis in field studies or even satellites (including the International Space Station; for example, vital information was relayed back to Earth from mass spectrometers on the Viking missions to Mars and from exploration of the atmosphere and frozen hydrocarbon surface of Titan, one of the moons of Saturn). A range of examples of other applications are described in the next sections.[1]

---

[1] See also, for example, the article: A. King, *Chemistry World*, Analytical Special, March 2014, I Mass spectrometry

Structural identification

Most mass spectrometers now store the collected spectra obtained in *El* and related experiments on organic molecules in a digital form. This has enabled large collections of spectra to be combined into spectral 'libraries' which can be rapidly examined to search for matches between an unknown compound's spectrum and those in the library. This facility which produces a rapid 'best fit' to the observed spectrum has dramatically assisted the identification of unknown samples even when only minute quantities are available for study—for example in the complex mixtures described later in this section.

This approach has very widespread applications, for example in areas of forensic analysis such as identification of illegal substances in samples taken from athletes.

A mass spectrometer also allows isotopes to be employed in mechanistic studies: for example, ^{18}O labelling can be used to determine whether hydrolysis of an ester, such as the reaction of H_2O with, e.g. $CH_3C(O)OCH_2CH_3$, proceeds with cleavage of the $CH_3C(O)-O$ or $CH_3C(O)O-CH_2CH_3$ bond.

Ionization energy and bond strengths

Attention can also be focused upon the processes which take place when electrons collide with molecules in the ionization chamber. For example, by studying the appearance of various peaks in a mass spectrum as the energy of the bombarding electrons is increased it is possible to determine the *ionization energy* of the molecule (i.e. the threshold energy at which the collision knocks out an electron) and also the *dissociation enthalpy* (energy) of a bond which is broken in a fragmentation process: as the ionization energy increases there is generally a greater degree of fragmentation, as higher vibrational levels are excited.

Quantitative applications

We have so far largely been concerned with the structural aspects of mass spectrometry, with emphasis on the use of wide scan ranges and determination of accurate masses and formulae. But the spectrometers also provide valuable information about the relative concentrations of these substances in a variety of applications— including drug concentrations, screening of metabolites in biochemical systems, clinical testing for disease biomarkers and environmental analysis of water samples. And smaller 'dedicated' mass spectrometers can give continual monitoring of the level of a chosen substrate or substrates (e.g. to give a real-time analysis of the products in a chemical reaction); medical examples include the quantitative analysis of the gases in human lungs during respiration.

Mass spectrometry and chromatography

We have demonstrated how the mass spectrometer provides an extremely sensitive technique for determining the molecular formulae and structures of a wide variety of organic compounds. Together with chromatography it can provide a rapid analysis of complicated mixtures of different compounds (typically for experiments involving relatively volatile material.)

A small quantity of the mixture to be examined is typically injected into a gas chromatograph (*GC*). The chromatograph separates the various volatile components (which are partitioned between the carrier gas stream and a liquid phase supported on an inert solid in the column) and these emerge from the chromatographic column in the carrier gas stream after different times have elapsed (depending on their retention times on the column). Each pure component is fed into the mass spectrometer (*MS*) and its spectrum is recorded with a rapid scan. In this way the *chromatograph* achieves the separation of components (the areas under the chromatographic traces indicate the relative amounts of the separate components in the mixture) and the *mass spectrometer* provides additional information which leads to the relative molecular masses and structures of the separate components. The combination of the two techniques (*GC/MS*) provides a remarkably sensitive and effective method for determining, for example, the products, and their relative yields, from a chemical reaction, and, as described below, is very effectively harnessed for investigation of complex mixtures from biological samples. An example involving the related gel permeation chromatography (*GPC*) is described in the next section.

Studies of biological molecules—applications in protein and peptide sequencing

Mass spectrometers are also increasingly employed to study biological molecules which, because of their size and (often) lack of volatility are not usually susceptible to study by electron impact and vaporization techniques. In one typical approach the positive ions from some examples can be produced, for example, by *Fast Atom Bombardment* or *laser desorption processes* (e.g. *MALDI*—see section 5.2): examples will be given which illustrate the challenges, as well as the results and interpretations of studying proteins and peptides, which tend to be involatile and not always water-soluble and often only available in minute quantities. Additionally, electrospray (*ESI*) techniques have also proved invaluable in generating positive ions for study.

Very complex fragmentation patterns are usually observed from biological samples. Then, one very helpful way to aid analysis is the use of *tandem mass spectrometry* (often referred to as *MS/MS*) in which ions of a given *m/z*, from the first mass spectrometer, are then passed directly into the focusing system of a second mass spectrometer; this usually leads to a simplified version of the spectrum which shows the fragmentation pattern from the single ion under investigation (rather than a complete set of fragmentations from many different ions).

We will illustrate the application of some of these techniques via the study of proteins and peptides, which are of course characterized by the peptide link which joins the constituent amino acids; i.e. –HN–CO–. Some of the amino acids commonly found are referred to in Table 5.1 which gives the name of each, and the commonly used code, as well as the structure of the key fragment (–NH–CHR–CO–) and its mass (referred to as the **residual mass**).

Table 5.1 Structures of the commonly occurring amino acids from proteins and peptides NH_2–CHR–CO_2H

Parent amino acid	Code	R	Residual mass (–NHCHRCO–)
Alanine (Ala)	A	$-CH_3$	71
Arginine (Arg)	R	$-(CH_2)_3NHC(=NH)NH_2$	156
Asparagine (Asn)	N	$-CH_2C(=O)NH_2$	114
Aspartic acid (Asp)	D	$-CH_2C(=O)OH$	115
Cysteine (Cys)	C	$-CH_2SH$	103
Glutamic acid (Glu)	E	$-(CH_2)_2C(=O)OH$	129
Glutamine (Gln)	Q	$-(CH_2)_2C(=O)NH_2$	128
Glycine (Gly)	G	$-H$	57
Histidine (His)	H	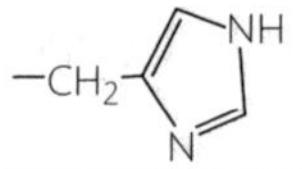	137
Isoleucine (Ileu)	I	$-CH(CH_3)CH_2CH_3$	113
Leucine (Leu)	L	$-CH_2CH(CH_3)_2$	113
Lysine (Lys)	K	$-(CH_2)_4NH_2$	128
Methionine (Met)	M	$-(CH_2)_2SMe$	131
Phenylalanine (Phe)	F	$-CH_2Ph$	147
Proline (Pro)*	P	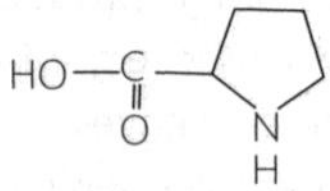	97
Serine (Ser)	S	$-CH_2OH$	87
Threonine (Thr)	T	$-CH(OH)CH_3$	101
Tryptophan (Trp)	W	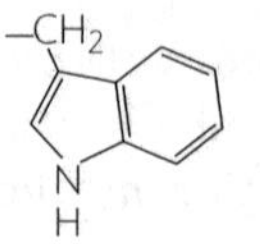	186
Tyrosine (Tyr)	Y	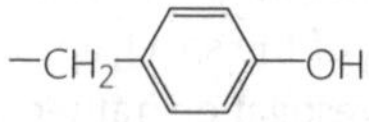	163
Valine (Val)	V	$-CH(CH_3)_2$	99

*Full amino acid structure

It is fortunate that the major pattern of fragmentation of peptides and proteins involves *cleavage of the individual peptide links*, to give, in each fragmentation, one or other, or both, of the two fragments which are best able to carry the positive charge (as described earlier for more simple organic molecules). This is illustrated here for a simple dipeptide, shown in Figure 5.19. The positive charge is associated with either the *carbonyl* group in one fragment or the *nitrogen* group in the fragment from the other

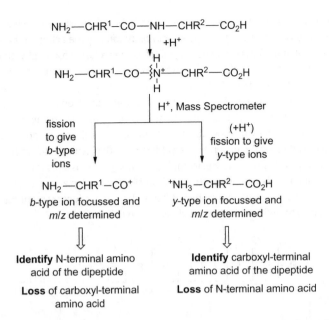

$$NH_2\text{—}CHR^1\text{—}CO\text{—}NH\text{—}CHR^2\text{—}CO_2H$$

$+H^+$

$$NH_2\text{—}CHR^1\text{—}CO\text{—}\overset{H}{\underset{H}{N^\pm}}\text{—}CHR^2\text{—}CO_2H$$

H^+, Mass Spectrometer

fission
to give
b-type
ions

$(+H^+)$
fission to give
y-type ions

$$NH_2\text{—}CHR^1\text{—}CO^+ \qquad\qquad {}^+NH_3\text{—}CHR^2\text{—}CO_2H$$

b-type ion focussed and
m/z determined

y-type ion focussed and
m/z determined

Identify N-terminal amino
acid of the dipeptide

Loss of carboxyl-terminal
amino acid

Identify carboxyl-terminal
amino acid of the dipeptide

Loss of N-terminal amino acid

Figure 5.19 Diagrammatic representation of the fragmentation of positively charged peptides in a mass spectrometer, showing the two major pathways (to give *b*- and *y*-type ions) for a model dipeptide.

mode of fission, as shown below. In the first case, referred to as fission to give a *b*-type ion (with charge on the carbonyl group), we will see a peak in the mass spectrum corresponding to the *m/z* of the nitrogen-terminal amino acid fragment (i.e. the amino acid less an oxygen atom) and in the second case, referred to as fragmentation to give a *y*-type ion (with charge on the nitrogen), the observed fragment will correspond to the terminal amino acid at the carboxyl end of the chain.

For a given peptide—say with a possible amino-acid sequence ABCDEF—a series of fragments will be seen identified as A^+, AB^+, ABC^+, $ABCD^+$, etc. depending on which bond is broken and/or F^+, EF^+, DEF^+, $CDEF^+$, etc. (i.e. for both *b*- and *y*- type fission respectively). These groups of fragments (e.g. A^+, AB^+, ABC^+) will differ in their molar masses by masses which correspond to the **residual masses** (of the amino acids B, C, D, etc.) shown in Table 5.1. Recognizing these differences, and identifying those formed by fission to *b*- or *y*-type ions, allows these related groups of ions, and hence their sequences to be determined, as illustrated later in this section.

In the examples typically studied, some of which are described here, there will be many amino acids in the peptide/protein chains, giving a bewildering complexity to the challenge of identifying the overall structure—which of course means determining the order of the individual amino acids. The simple guidance given above, and some perseverance, will help. Increasingly, detailed computer analysis can help identify the possible combination of amino acids which account for each of the myriad peaks, and hence reveal the complete structure.

Worked examples in peptide sequencing

(i) We will examine first the mass spectrum shown in Figure 5.20 obtained by the *ES-MS/MS* technique, of an octapeptide, *angiotensin II*. This is a peptide hormone that causes vasoconstriction and an increase in blood pressure in the human body. It is produced from the precursor *angiotensin I* by the action of the enzyme *ACE* (*Angiotensin Converting Enzyme*) which is then a target for inactivation by *ACE* inhibitor drugs which are used in the treatment of hypertension. The mass spectrum shown in Figure 5.20 also shows the relationship between some of the main peaks, marked to indicate the two major fragmentation types (to give *b*- and *y*-type ions). Also note that the molecular ion, from protonated angiotensin II does not appear here (presumably fragmentation is particularly ready), but can be measured in related experiments as MH$^+$ *m/z* 1046, which is the molecular weight (and corresponds to the formula $C_{50}H_{71}N_{13}O_{12}$ plus a proton). In tackling this, potentially very complex, structural problem, we will place great emphasis on picking out series of peaks which differ in their *m/z* values by numbers which correspond to individual amino acids as shown in Table 5.1.

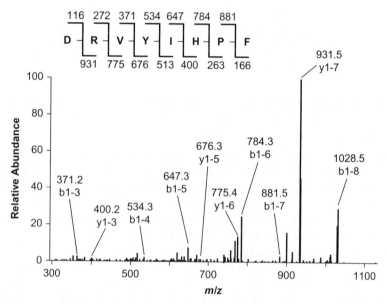

Figure 5.20 Mass spectrum of angiotensin II (MH$^+$ *m/z* 1046) obtained by *ES-MS/MS* experiments. Reprinted with permission from E. Buyukpamukcu, D. M. Goodall, C-E. Hansen, B. J. Keely, S. Kochhar, and H. Wille, *J. Agric. Food Chem*, 2001, **49**, 5822. Copyright (2001) American Chemical Society.

Let's first consider as our starting point the peak at *m/z* 1028 (omitting the decimal point), which corresponds to the loss of 18 mass units from the molecular ion; we can identify this as loss of the elements of water, to leave the chain with the charge on the carbonyl group (RCO$^+$) at the 'right-hand' end of the peptide chain (i.e. a *b*-type ion). To identify the first amino acid in the chain, we look for a peak associated with *b*-type fission (i.e. loss of the first amino acid from the carboxyl end), which differs from 1028 by a mass equivalent to one of the residual masses in Table 5.1. The peak at 881 indicates a 'loss' (i.e. mass difference) of 147 compared to the peak of 1028, which corresponds to the amino acid *phenylalanine*, which suggests this amino acid as the carboxyl-group terminus. From *m/z* 881 to the peak at *m/z* 784 corresponds to a mass difference of 97, which characterizes the amino acid *proline* as the next amino acid in the series: i.e. the peak at 784 corresponds to the loss of a total of 244 mass units from 1028 (that is 147 + 97). Next, compared to the peak of 784, is the 'loss' (mass difference) of 137 mass units, to give the peak at 647, which characterizes *histidine* as the next amino acid in the chain. The next 'loss' is 113, down to 534, indicating *iso leucine* as the next amino acid (separate experiments involving detailed analysis of fragmentation patterns can be used to distinguish this from the alternative *leucine*, with the same relative molar mass). You are encouraged to try to follow the next key mass difference down the series of *b*-type fragmentations to continue the chain and check these with the solution given in Figure 5.20.

Similarly, we can identify the peak at *m/z* 931 as being a difference of 115 mass units from the molecular ion ('mass loss'), and hence the identification of the amino acid *aspartic acid* as that at the NH terminal end of the chain, and formed by *y*-type fission; the peaks at 775 and then 676, and 513, correspond to the mass differences of 156 (*arginine*), then 99 (*valine*), then 163 (*tyrosine*), respectively. Why not try to confirm the next amino acids in this series? These results should correspond to those obtained by analysis of the *b*-type fragmentation (i.e. working from the other end) as explained above.

Putting all the information together, and double-checking since we can get information from both modes of fission, gives the following structure for angiotensin II;

asp – arg – val – tyr – ileu – his – pro – phe
(amino terminus)   (carboxyl terminus)

Note that the full structure is given on the Royal Society of Chemistry's ChemSpider database (see also Online Resource): see also page 97, section 4.8 for reference to its NMR spectra.

(ii) This example is also taken from the research paper illustrated above (and referred to in Figure 5.20), which also illustrates the sensitivity of the technique and its use in conjunction with chromatographic methods of analysis of mixtures.

It is known that the unique flavour of chocolate derives from the processing of cocoa beans via fermentation, drying and roasting. It is believed that fermentation leads to the breakdown of cocoa proteins into smaller peptides which then react with sugars to give the flavour components. In this particular study the cocoa beans, and their proteins were degraded by fermentation, followed by separation of the resultant polypeptides by *GPC* (*gel permeation chromatography*) and *mass spectrometry*, in particular by the *EPI-MS/MS* technique. The ultimate aim was to identify the structure of the proteins responsible for the peptides which may themselves be sources of flavour components by determination of the number and order of the constituent amino acids. (Note also that if in a series of experiments of this sort, a number of peptides can be obtained, studied (e.g. via selective enzymic cleavage) and their structure and 'overlaps' can be identified, then the whole chain sequence of the parent protein can in principle be identified, much as Frederick Sanger and his colleagues did for their classic study using paper chromatography to reveal the arrangement (sequencing) of the 51 amino acids which characterize the insulin hormone.)

One particular *hexapeptide* has the mass spectrum shown in Figure 5.21, with a series of peaks identified which are believed to be due to the two types of fragmentation (i.e. from the carboxyl and amino ends), as shown in the earlier example. The parent ion has *m/z* 621.3 (for MH⁺). With this information, can you identify the amino acids lost in each type of fragmentation, as far as possible, by comparison with the information on residual masses in the Table, and hence combine information on both types of fission to suggest a structure?

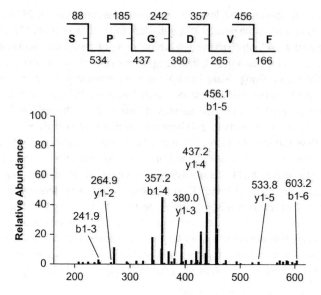

Figure 5.21 Mass spectrum derived by tandem *MS* (*MS/MS*) study of the protonated molecular ion (MH⁺ *m/z* 621.3) obtained from a hexapeptide derived from the cocoa vicilin protein (see legend for Figure 5.20). Reprinted with permission from E. Buyukpamukcu, D. M. Goodall, C-E. Hansen, B. J. Keely, S. Kochhar, and H. Wille, *J. Agric. Food Chem*, 2001, **49**, 5822. Copyright (2001) American Chemical Society.

Let's first examine the series of peaks from *b*-type ions, starting with the loss of the first fragment to give the peak with *m/z* 603 derived from the protonated molecular ion at 621; this is simply loss of water from the carboxyl terminus of the protonated parent to leave a fragment RCO⁺. The next relevant fragment ion is 456, i.e. 'loss' (mass difference) of 147, which characterizes *phenylalanine*; next comes a mass difference of 99, to give the peak at 357 (i.e. loss of *valine*); the subsequent 'loss' of 115 (to 241.9) suggests that *aspartic acid* is next. Working from the amino terminus, we note the first peak at *m/z* 533.8, i.e. M–87, which characterizes *serine*, the 'loss' of 97 (*proline*) and then 57 (*glycine*) (peaks with *m/z* 437 and 380 respectively). The next fragmentation has a mass difference of 115 (*aspartic acid*) (to give the peak at *m/z* 264.9), which confirms the latter's position in the chain, and leads to the following structure for the hexapeptide:

ser – pro – gly – asp – val – phe

Information on a related nonapeptide in the mixture is provided for you to solve in Exercise 5.4 at the end of the chapter.

5.7 Summary

As a result of studying the material presented in this chapter, you should have developed an understanding of the basic principles which underpin the technique of *mass spectrometry* and its application to structure determination.

These key features include:

- the ways in which *positive particles* can be generated from molecules and their charge-to-mass ratio (*m/z*) determined (with an appreciation of the basic principles of instrumentation design and usage);

- the determination of *relative molar masses* (molecular weights), as well as isotopic compositions and formulae, all from the molecular ion;
- the principles which underlie the *fragmentation* patterns observed and the clues which can hence be obtained about the *molecular structures* of compounds.

You should also now have *problem-solving skills* in the use of mass spectrometry (which can be used together with other spectroscopic techniques like NMR) for solving structural problems.

You should also have developed an appreciation of the extreme *sensitivity* of mass spectrometry, in terms of the very small amounts of materials needed, and the wide range of applications of this technique in chemistry and biology. These include the widespread use of *MS-GLC* techniques for analysis of complex mixtures, e.g. in analysing reaction products and solving analytical problems, and the steadily increasing uses in tackling problems in biological chemistry.

5.8 **Exercises**

5.1 Identify the compound whose mass spectrum is shown in Figure 5.22. The peak at m/z 61 is approximately 2% of the height of that at m/z 60. Give the molecular formula and calculate the accurate molar mass.

Worked solutions to the exercises are available on the Online Resource

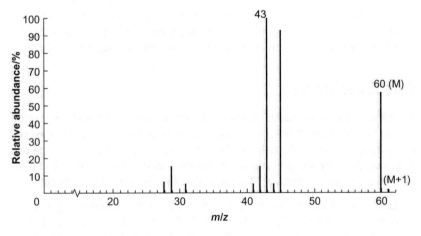

Figure 5.22 Mass spectrum for Exercise 5.1.

5.2 Sketch the relative intensities and m/z values of the molecular ions of different isotopic composition in the mass spectra of (a) 1,1-dibromo-ethane, and (b) dichloromethane.

5.3 Identify the compound (of suggested molecular formula $C_9H_{10}O_2$) whose mass spectrum is shown in Figure 5.23. The ^1H spectrum is described in Exercise 4.4, page 101 and the IR spectrum in Exercise 2.13, page 51. How do you account for the formation of a fragment with m/z 108?

5.4 We have discussed in section 5.6 the separation and identification of peptides believed to be in part responsible for the flavour of chocolate, and identified

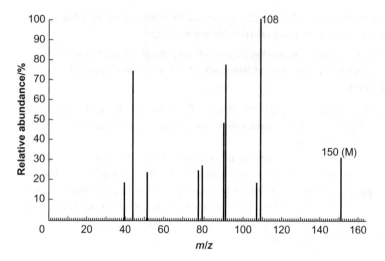

Figure 5.23 Mass spectrum for Exercise 5.3.

an unknown hexapeptide (as a worked example). This study also produces a *nonapeptide*, with relative molar mass 902.5 (for MH⁺), with a series of peaks as follows; 902.5, 884.3, 831.3, 737.2, 734.4, 638.2, 621.1, 534.1, 523.3, 466.3, 437.2, 369.1. It is not clear at this stage which of these belongs to which series (i.e. b- or y-type). Work out which of these belongs to which series and identify the appropriate masses, using the residual masses of the amino acids given in Table 5.1. Hence work out the order of the amino acids and the structure as a whole.

Further exercises

The following two problems illustrate how the combination of techniques we have discussed so far (IR, UV, and NMR spectroscopy, with mass spectrometry) can be employed to solve organic structural problems. The information is supplied here in condensed and/or tabulated form, which is typical of the presentation of much summarized spectra in research papers, especially where several compounds are to be described.

5.5 Identify the structure of the compound X for which the C,H,N analysis gives a composition 62.0% C, 27.6% O, and 10.4% H. The UV spectrum of X, recorded for a solution in hexane, has λ_{max} 290 nm, $\varepsilon = 1.8$ m^2 mol^{-1} and the IR spectrum shows strong peaks at 2950, 2700, 1720, and *ca.* 1400 cm^{-1}.

The ten most abundant peaks in the mass spectrum are as follows:

m/z	29	58	28	27	57	18	41	39	15	55
relative abundance/%	100	83	82	57	26	8	7	6	5	4

The ^{1}H NMR spectrum has resonances (δ/ppm) at 1.05 (3H, triplet $J = 7$ Hz), 2.45 (2H, quartet $J = 7$Hz, doublet $J = 2$ Hz) and 9.86 (1H, triplet, $J = 2$ Hz). The fully decoupled ^{13}C NMR spectrum has three resonances at δ 204, 38 and 7.

5.6 Identify the compound Y, which has the composition 90.5% C and 9.5% H. The UV absorption spectrum has a very strong absorption at *ca.* 200 nm, and an absorption (with vibrational fine structure) at *ca.* 260 nm, and a much weaker absorption in the range 300–350 nm. The IR spectrum has strong bands at 3050 and 2950 cm^{-1} as well as at *ca.* 750 cm^{-1}.

The most abundant peaks in the mass spectrum are as follows:

m/e	91	106	51	39	65	77	92	78	27	105
relative abundance/%	100	31	13	10	8	8	8	7	6	6

The ^1H NMR spectrum has three major resonances (δ/ppm) as follows: 7.1 (5H, complex), 2.50 (2H, quartet, $J = 7$ Hz, 1.10 (3H, triplet $J = 7$ Hz).

The fully decoupled ^{13}C NMR spectrum has resonances at δ/ppm values of 144.2, 128.3, 127.9, 125.7, 28.9, and 15.6.

5.9 **Further reading**

L. M. Harwood and T. D. W. Claridge (2015), *Introduction to Organic Spectroscopy, 2nd Edition*, Oxford Chemistry Primers, Oxford University Press, Oxford.

6 X-ray diffraction and related methods

6.1 Introduction

The diffraction techniques we describe here are quite distinct from the spectroscopic (and spectrometric) methods so far discussed. Thus, whereas the spectroscopies are based on the absorption of certain wavelengths (and energies) from radiation with a range of wavelengths, diffraction techniques employ radiation (e.g. X-rays) with a single wavelength, which we refer to as *monochromatic* radiation. X-ray diffraction, for example, occurs when a monochromatic beam of X-radiation interacts with matter and is *scattered* in different directions, with no absorption of energy. Similarly, beams of neutrons or electrons with well-defined wavelengths can be scattered to give typical diffraction patterns.

The basis of the application of diffraction techniques in solving chemical problems is then to use ions or molecules as diffraction gratings and to determine, from the observed diffraction phenomena, the spacings between ions in a crystal or between the atoms in molecules.

This chapter will first provide an introduction to the principles of X-ray diffraction: the experimental design and the interpretation, via the use of Bragg's Equation, of the diffraction patterns produced by single crystals and ionic lattices. Analysis to give the unit-cell type, and dimensions, will be provided, along with opportunities to work through related problems. This will then take us to consideration of the information which can be obtained about molecules, including some complex examples of particular biological relevance (for example with electron density maps and determination of full molecular coordinates). In the last section, a brief explanation of the main features of neutron and electron diffraction and their application to structural problems will be provided.

6.2 Introduction to the X-ray diffraction method

The first significant experiments were carried out at the beginning of the 20th century when it was realized both that X-rays have wave properties and that crystals consist of regular arrays of atoms or ions. It was demonstrated by von Laue that a crystal lattice behaves as a grating, so that it is possible to generate a diffraction pattern from a crystal; for his first experiments he used an ionic crystal, a beam of X-rays (with a range of wavelengths) and a photographic plate (to detect the scattered X-rays). A regular pattern of spots appeared on the plate, giving a clear indication of the success of the experiment. As we shall see, diffraction phenomena can be observed if the wavelength of the radiation is of the same order of magnitude as the 'repeat distance' between the atoms or ions in a crystal; X-rays (but not visible light) fulfil this condition.

The method has been developed to provide a means for determining the exact positions of ions in a crystal lattice and of atoms within a molecule—that is, for determining accurate values for bond angles and bond lengths, even in extremely complicated molecules like proteins and enzymes.

The apparatus

X-rays are produced when a beam of accelerated electrons strikes a metal target. An inner electron from an atom in the metal is ejected, an outer electron drops down to fill the vacancy created, and the emitted radiation, of precise energy, frequency (v), and hence wavelength (λ), is in the X-ray region. Since various electronic transitions are possible, the resultant beam at this stage contains X-rays of several different energies (wavelengths). Figure 6.1 for example, shows the intensity of X-radiation, as a function of λ, emitted from a copper target. The K_α line, which corresponds to the energy emitted when an electron undergoes a transition from the L shell to the K shell ($2p \rightarrow 1s$ in terms of orbitals), has $\lambda = 0.154$ nm. A sheet of nickel proves to be a good filter for all the wavelengths except the K_α line (all the X-rays of wavelength less than the 'absorption edge' of 0.149 nm are absorbed: they are of high enough energy to remove completely a K-shell electron from a nickel atom) so the combination of the two metals used like this provides a *monochromatic beam* of radiation.

The monochromatic beam of X-rays is then incident on a solid sample (single crystal or powder) of the material under investigation. A powder contains many very small crystals, in a variety of different orientations, whereas for a single crystal only one orientation of the solid can be considered at a time.

Detection of the resultant X-ray beam was typically achieved by surrounding the sample with a photographic film: where X-rays strike the film it became darkened, and the film was subsequently developed to yield the diffraction pattern. Figures 6.2a and 6.2b show the experimental arrangement for taking a powder 'photograph'.

Figures 6.3 and 6.4 show the thin strips of film (opened out) recording the diffraction patterns from two powdered metals (molybdenum and copper, respectively). Figure 6.5 is the diffraction pattern from powdered sodium chloride. The photographs

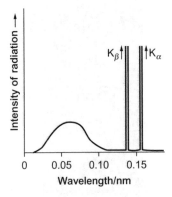

Figure 6.1 Intensity of X-radiation at different wavelengths emitted from a copper target.

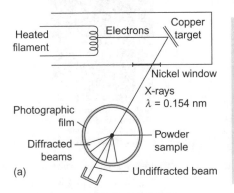

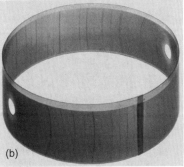

Figure 6.2a Basic features of an X-ray camera for use with powdered samples.
Figure 6.2b Arrangement of the film for a powder photograph: the resulting 'lines' can be seen.

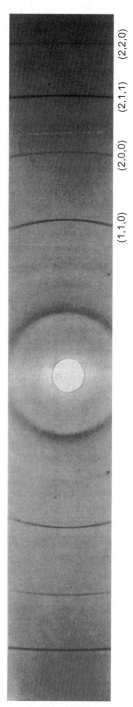

Figure 6.3 X-ray diffraction photograph from a powder sample of molybdenum.

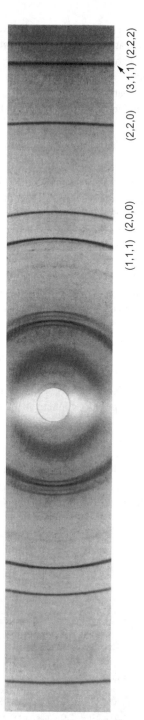

Figure 6.4 X-ray diffraction photograph from a powder sample of copper.

Figure 6.5 X-ray diffraction photograph from a powder sample of sodium chloride.

indicate that the X-rays, on striking the powders, become diffracted into a series of well-defined cones. The origin of this phenomenon will be discussed in the next section.

If a *single crystal* is employed, it is usually mounted at the centre of a cylindrical film of somewhat greater depth than that used in the powder method. The crystal is arranged with one of its major axes vertical. The diffraction pattern is then recorded, often with simultaneous rotation of the crystal about the axis. A typical *single crystal photograph* shows several *layers* of spots; Figure 6.6 is an X-ray single crystal rotation photograph of sodium chloride. The derivation of information from the diffraction patterns will follow the explanation of the principles which underpin the observations, which will be explained in the next section.

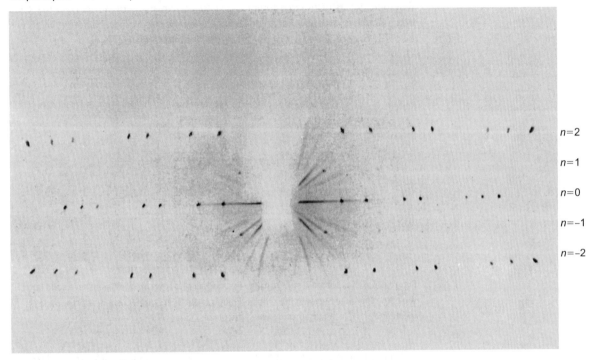

Figure 6.6 Single crystal rotation X-ray photograph (from sodium chloride).

In modern X-ray diffractometers, the instrumentation has been significantly developed, for example using *CCDs* (*charge-coupled devices*), based on technology related to that employed in digital cameras, to detect diffracted X-rays, with computer control of data collection and presentation. In this way, vast amounts of information can be rapidly collected, analysed and displayed in ways which can reveal the structures of highly complex molecules: these include many of major biological interest, for example proteins and enzymes, as explained in section 6.4, and for which X-ray beams from a synchrotron may be employed.

The Bragg Equation

A more detailed understanding of X-ray diffraction and of the exact requirements for the appearance of intensity maxima were presented by W. L. Bragg in 1912 (this proposal was based substantially on closely related and similarly pioneering work by his father, W. H. Bragg, both of whom were awarded Nobel Prizes[1]). He realized that when X-rays impinge on a crystal, some are reflected from the atoms in the top layer, whereas

[1] An excellent short summary of the Braggs' work is given in the account by A. Sella, *Chemistry World*, December 2013, p 37.

others penetrate this layer and are reflected off the next layer, and so on. Analysis shows that the resultant reflected rays are only in phase for certain angles of incidence of the X-rays upon the crystal. This is illustrated in Figure 6.7, which shows the path difference between the two rays (one reflected off the top layer, the other reflected off the second layer) when they arrive at the at the detector. For these two rays to combine (i.e. to reinforce each other) they must be completely in phase: that is, their path difference ($2d\sin\theta$) must be a whole number of wavelengths ($n\lambda$) where d is the separation between the planes and λ is the wavelength of the X-rays. This is the **Bragg Equation** (Equation (6.1)) and it correctly predicts that a reflected beam will be observed (i.e. that there is *constructive* interference) only for certain angles of incidence of the X-ray beam on the crystal; at other angles of incidence the rays from the different layers will be partly or completely out of phase *(destructive* interference).

$$n\lambda = 2d\sin\theta \qquad\qquad (6.1)$$

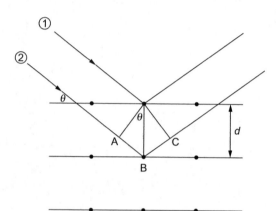

Figure 6.7 Reflection of X-rays from the first and second rows of atoms (ions) in a solid lattice; the path difference between the reflected rays is $(AB+BC)=2d\sin\theta$, where θ is the angle of incidence and d is the distance between the planes.

For example, if $\lambda = 0.154$ nm and $d = 0.2$ nm, then as θ (the angle of incidence) is steadily increased, reflection will first occur when $n = 1$ and $\sin\theta = 0.385$, i.e. when θ is approximately 23°. The condition may also be satisfied for $n = 2,3,4$, etc., and hence for higher values of θ, but we will normally be concerned with the *first order* ($n = 1$) reflections.

Figure 6.8 illustrates one particular small crystal (crystallite) in a *powder*, orientated with its surface plane of atoms at an angle θ to the beam. If this value of θ fulfils the

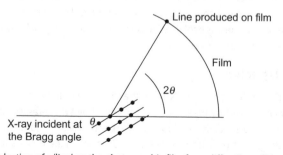

Figure 6.8 Production of a 'line' on the photographic film from diffraction of X-rays incident at the Bragg angle.

Bragg Equation for the particular values of λ and θ in the experiment (since there are many crystallites in the powder, this will be true for some of them), then the beam is reflected on to the film. There will be other crystallites each making the angle shown, which means that a *cone* of diffracted X-radiation will be produced, darkening the film at the points indicated (see also Figures 6.2–6.5). A *series* of cones is produced because there are various possible spacings in the crystal (with different values of *d*) which satisfy Equation (6.1) for different values of θ.

When a *single crystal* is used, with one axis vertical, then the regular interatomic spacing along this axis behaves as a simple diffraction grating (Figure 6.9). Thus, the layers of lines observed (see Figure 6.6) are simply the reflections which satisfy Equation (6.2) for λ, *d* (the vertical spacing, Figure 6.9) and *n* = 1,2, etc.

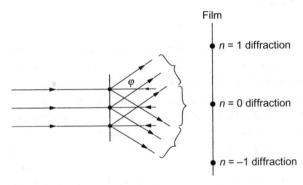

Film

● *n* = 1 diffraction

● *n* = 0 diffraction

● *n* = −1 diffraction

Figure 6.9 Diagrammatic representation of diffraction by a single crystal mounted with one axis vertical; the diffractions with *n* = 1,0, etc. correspond to the layers of lines clearly visible in Figure 6.6.

The layer lines themselves are clearly not continuous, because for some points on the layer lines there will be destructive interference between the reflections which satisfy Equation (6.2) for the vertical spacing and reflections from other planes in the crystal. To understand this and to appreciate the relationship between the possible spacings in a crystal (and hence all the possible values of *d*) and the structure of the lattice, some familiarity with crystal types is needed, as described in the next section.

$$n\lambda = d\sin\varphi \qquad (6.2)$$

6.3 Crystallography

Unit cells and crystal systems

A crystal consists of a repeating **unit cell** of atoms or ions, in three dimensions. The unit cell is characterized by the length of each side (*a*, *b*, *c*) and the angles between the three sides (*α*, *β*, *γ*). Figure 6.10 shows two of the possible different types of unit cell or **crystal system**. The symmetry of each of these allows three-dimensional structures (crystals) to be built up from the tiny building blocks (unit cells). The planes referred to earlier are sheets in the crystal containing a high density of lattice points (atoms or ions), observed as the external faces of a crystal.

Within the unit cell there are various allowed arrangements of atoms or ions which still preserve the overall symmetry. Figure 6.11 shows the three possibilities for a **cubic**

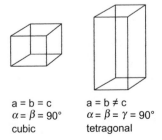

a = b = c
α = β = 90°
cubic

a = b ≠ c
α = β = γ = 90°
tetragonal

Figure 6.10 Two crystal systems.

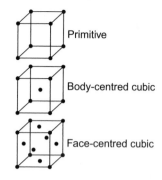

Figure 6.11 The three possible cubic unit cell arrangements.

unit cell (i.e. with equal sides, all angles 90°); these are the **primitive** (or **simple**) cubic, **body-centered** cubic (bcc), and **face-centered** cubic (fcc) systems: these are called Bravais lattices.

You will probably find it helpful to inspect 'ball-and-stick' models of unit cells and fragments of lattices in order to appreciate the various possible planes of atoms or ions in the different types of crystal. You may also find it helpful that a face-centred cubic lattice can equally well be described as a cubic-close-packed arrangement; this common structure arises if equivalent spheres are packed in one layer, the next layer is added, and then the third layer is added at positions not corresponding to those of the first layer, but in the alternative positions; the next layer corresponds to the first, and the pattern is repeated.

Lattices in solids

The lattices to be studied may be of several types.

(i)   *Atoms (metals or alloys).* For example, metallic caesium exists in a body-centered cubic lattice, as does α-iron, whereas copper has a face-centered cubic pattern. There is only one type of atom in the unit cell.

(ii) *Ionic crystals.* These contain two or more different types of ion (e.g. caesium chloride, sodium chloride), and they can be characterized in the same way as the simple lattices. For example, caesium chloride consists of a simple cubic array of Cs^+ ions with Cl^- ions at the body centre and *vice versa*. The arrangement is referred to as two inter-penetrating primitive cubic lattices. For sodium chloride the crystal structure is of two inter-penetrating face-centred cubic lattices: the complete unit cell is shown in Figure 6.12. Remember that

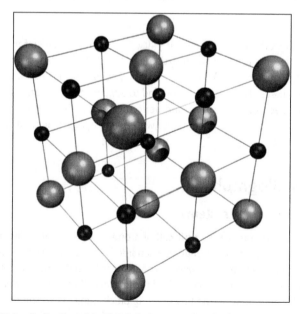

Figure 6.12 Unit cell of sodium chloride, NaCl. One type of ion (Na^+ or Cl^-) can be seen at the corners of the cube and at the centre of each face: the other ion occupies vacancies created in this lattice and itself has a face-centred cubic structure. Reproduced via the CC BY 3.0 licence, © Ausis.

the use of small balls and 'bonds' for models is essentially for convenience in visualizing the planes and unit cells. Close packing of larger space-filling spheres produces a more realistic model in which the electron clouds of neighbouring ions are seen to be in contact.

(iii) *Covalent molecules*. Covalent compounds also form crystals, with *molecules* at the lattice points in the unit cells, as for atoms and ions.

Planes in the crystal

A shorthand procedure, which is known as the use of **Miller Indices**, is used for describing a particular plane in a crystal. Their definition can be illustrated for the three-dimensional lattice illustrated in cross-section in Figure 6.13; the x and y axes are indicated, with the z-axis coming out of the paper. Each dot then represents a vertical column of atoms. The three lines indicated are three planes which we wish to describe.

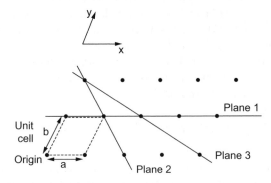

Figure 6.13 Representation of different planes within a crystal.

The procedure, summarized in Table 6.1, is as follows:

(i)   choose an atom as origin;

(ii) read off the intercepts along the x, y and z axes in terms of the unit spacings in the crystal (a along the x-axis, b along the y-axis, c along the z-axis); the intercept is infinity if the plane does not cut the axis concerned;

(iii) take reciprocals, dropping any reference to a, b, and c; and

(iv) if the reciprocals for a given plane include a fraction or fractions, express the reciprocals as their simple integral ratio, leaving **Miller Indices**, which are then referred to as h,k,l values.

Table 6.1 Derivation of Miller Indices (h,k,l) for 3 planes shown in Figure 6.13

Step (i) Plane	(ii) Intercepts	(iii) Reciprocals	(iv) Miller Indices h,k,l
(1)	$\infty a, 1b, \infty c$	$1/\infty, 1/1, 1/\infty$	(0,1,0)
(2)	$2a, 2b, \infty c$	$1/2, 1/2, 1/\infty$	(1,1,0)
(3)	$4a, 2b, \infty c$	$1/4, 1/2, 1/\infty$	(1,2,0)

Thus plane (1) and *all those parallel to it* are referred to as (0,1,0) planes; similarly plane (2) and those parallel to it are called (1,1,0) planes. You should check that it does not matter where the origin is taken.

To recapitulate: *a*, *b*, and *c* are properties of the unit cell, as are the angles between axes; *h*, *k* and *l* are integers used to describe any particular plane in the crystal and, as will be seen, they are used to calculate the distance between the planes (i.e. *d* in the Bragg Equation). Incidentally, a three-dimensional lattice can cleave (or grow) along any of these planes, which accounts for the external forms of crystalline compounds.

Worked examples using Miller Indices

At this stage you are encouraged to classify some planes to become familiar with the nomenclature. Given the planes and axis system in Figure 6.14, (i) what are the Miller Indices for each plane, and (ii) can you draw the (0,0,2) plane?

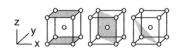

Figure 6.14

To visualize a plane running through the crystal it is perhaps best to consider several unit cells together, or even better, to refer to a model. The answers for the planes shown are from left to right, (0,0,1), (1,1,0), and (1,1,1), respectively. Note that the (1,0,0) and (0,1,0) planes simply relate to the (0,0,1) plane, i.e. they are all the appropriate 'ends' of the cube. Similarly, (1,0,1) and (0,1,1) will be diagonal planes (like 1,1,0).

The (0,0,2) planes are as shown in Figure 6.15. To check this, note that the intercepts are ∞a, ∞b, 1/2 c, with *reciprocals* (0,0,2).

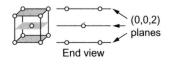

Figure 6.15

Linking the Bragg Equation and Miller Indices (*h,k,l*)

It is often necessary to describe the perpendicular distance d_{hkl} between parallel planes (*h,k,l*) of a unit cell with lengths *a*, *b*, and *c*. This is the inter-planar spacing referred to in the Bragg Equation. With some relatively simple geometry it can be shown that for an **orthogonal** crystal system (a unit cell with all angles 90°):

$$\frac{1}{d_{hkl}^2} = \frac{h^2}{a^2} + \frac{k^2}{b^2} + \frac{l^2}{c^2} \tag{6.3}$$

and for a **cubic** system:

$$d_{hkl} = a / \sqrt{\left(h^2 + k^2 + l^2\right)} \tag{6.4}$$

We can use this to calculate *d* for a given set of planes; for example, the distance between the (0,0,2) planes in a cubic system is *a*/2, as expected from Figure 6.15.

6.4 Determination of structure

Unit cell type and dimensions

X-ray data from a given ionic crystal—in terms of the particular value of θ at which reflections are observed—can be used to determine the type of unit cell in the compound. The Bragg Equation (for the allowed values of θ (angle of incidence) for given λ and *d*) is combined with the expression for the possible values of *d* in terms of the cell dimensions (*a,b,c*) of the particular structure considered. For a cubic lattice, for which the spacing d_{hkl} between any set of planes (*h,k,l*) is given by Equation (6.4), the resulting expression which gives the angles of reflection from different planes is derived as shown, where *a* is the length of the cell side (Equation (6.5).

$$n\lambda = 2d\sin\theta \tag{6.1}$$

$$\therefore \sin^2\theta = \frac{(n^2\lambda^2)}{4d^2}$$

$$\therefore \sin^2\theta = \frac{(n^2\lambda^2)}{4a^2}\left(h^2 + k^2 + l^2\right) \tag{6.5}$$

There are many lines in the diffraction pattern (i.e. at different values of θ) owing to the various possible values of d. Since, for any set of planes, h, k, and l must all be integers, then so must $(h^2 + k^2 + l^2)$; the possible values of h, k, l, and $(h^2 + k^2 + l^2)$ are shown in Table 6.2.

Table 6.2 Values of $(h^2 + k^2 + l^2)$ for different planes h,k,l

h,k,l	1,0,0	1,1,0	1,1,1	2,0,0	2,1,0	2,1,1	2,2,0	2,2,1 / 3,0,0	3,1,0
$(h^2 + k^2 + l^2)$	1	2	3	4	5	6	8	9	10

Thus for a cubic lattice, the increasing values of $\sin^2\theta$ should be related to each other as are the increasing simple integers 1,2,3,4, etc (the smallest value of $\sin^2\theta$ being the reflection from the (1,0,0) planes, the next from the (1,1,0) planes, etc.) but with no line corresponding to the integer 7 because no combination of the squares of integers gives this number. Similarly it can be shown that there should be no lines for sums of integers 15, 23, 28. The pattern obtained from a simple cubic lattice confirms this analysis, and any solid which gives such a pattern (i.e. with 7th, 15th, etc., lines missing) is known to have a cubic lattice. Further, if λ is known, values of $\sin^2\theta$ for the $n = 1$ reflections lead to a measurement of a, the length of the side of the unit cell. The process just described is called **indexing** a powder photograph (that is, deriving the shape and size of the unit cell).

In the single crystal method, measurements with each axis vertical in turn lead to determination of the unit cell length along each of the axes (from the separation of the layer lines, see page 135, section 6.2). Again, comparison with patterns from compounds whose structures are known can be helpful.

The Bravais lattice

The type of Bravais lattice (e.g. body- or face-centred) can also be readily obtained from the diffraction pattern. First, consider a body-centred cubic structure and try to visualize what happens to the expected reflections from the (1,0,0) (and the (0,0,1), (0,1,0) planes). For the (1,0,0) planes viewed end-on, it can be seen that the reflection which satisfies the Bragg Equation (i.e. with distance d_{100}) now has an extra beam superimposed (Figure 6.16). This is the beam reflected from the atoms at the centres of the unit cells; as shown this extra beam is completely *out of phase* with the beams reflected from the (1,0,0) planes (the path difference for the extra beam compared to the others is $\lambda/2$ and therefore there will be destructive interference; the reflection is now absent). This is referred to as a **systematic absence**.

The (2,0,0), (0,2,0), and (0,0,2) reflections (with separation d_{200}) will be present, the reflections being in phase (remember that d is now half that for the (1,0,0) planes); a diffracted beam at the corresponding θ is observed. As Figure 6.17 demonstrates, the (1,1,0) reflection should be *present* (there being no atoms in-between the (1,1,0) planes) but the (1,1,1) plane reflections are absent (the planes have atoms in between).

For a *face-centred* cubic structure, the (1,0,0) reflections are absent, as too are the (1,1,0) reflections, but not, in this case, the (1,1,1) type (you are recommended to check that this is the case by reference to models or 3D sketches). Table 6.3 summarizes the observed reflections for different types of cubic crystal.

Figure 6.16 A body-centred cubic structure.

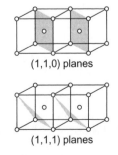

(1,1,0) planes

(1,1,1) planes

Figure 6.17 A body-centred cubic structure showing planes with interstitial ions (1,1,1) and planes without interstitial ions (1,1,0).

Table 6.3 Allowed reflections for cubic crystals

Cell type	(h,k,l)	1,0,0	1,1,0	1,1,1	2,0,0	2,1,0	2,1,1	2,2,0
Primitive cubic		√	√	√	√	√	√	√
Body-centred cubic			√		√		√	√
Face-centred cubic				√	√			√

We can summarize by noting that a simple (primitive) cubic cell gives no systematic absences, whereas the body-centred cubic cell shows only those reflections for which the sum $(h + k + l)$ is even; the face-centred cubic structure gives reflections only from planes where h, k, and l are either all odd or all even. This information enables you to use a series of values of $\sin^2\theta$ (from a powder photograph) to decide from which type of lattice these values arise. For example, if the values of θ are such that the ratios of $\sin^2\theta$ for the first three reflections fit most closely the integers 3, 4, and 8, then the lattice is of *face-centred cubic* type (3 has h,k,l, of (1,1,1), 4 has h,k,l of (2,0,0,), and 8 has h,k,l (2,2,0)): check that the next reflections will have sums of the squares of the integers equal to 11 and 12. The ratios are quite different for primitive or body-centred cubic structures.

This procedure can be exemplified by reference to Figures 6.3 and 6.4; you should be able to conclude by inspection at this stage that molybdenum has a body-centred cubic structure, whereas copper is face-centred cubic. Note that the photographs are reproduced to scale (180 mm = 180°), so that you should also be able to verify that the appropriate values of θ (measured from the film—the angle between a line and the centre of the film is 2θ) give integral ratios of $\sin^2\theta$ as predicted. Worked examples and Exercises are given later in the chapter.

Ionic crystals

The powder photograph of crystalline sodium chloride (page 132, Figure 6.5) provides a good example of the effect upon the diffraction pattern when more than one type of ion is present in the lattice. It can be seen from the diffraction pattern that the (1,1,1) reflections are weaker than those for the (2,0,0) planes. This observation can be understood in terms of the intensity of the scattered beam of X-rays and its dependence on the nature of the atom doing the scattering. The X-rays are scattered by the electrons around the nucleus, which will lead to a different intensity (amount) of scattering from Na^+ and Cl^-. Now, for the (2,0,0) type reflections, all the planes causing reflection contain Cl^- and Na^+ ions so that all these ions will contribute to reinforcement (three (2,0,0) horizontal layers are clearly shown in Figure 6.12). However, for the (1,1,1) plane of sodium ions, there is a layer of chloride ions in between (see Figure 6.18); the X-rays scattered from each layer are exactly out of phase, but do not cancel exactly because their *intensity* is not the same. Thus a weak reflection is observed.

For KCl the two ions have the same number of electrons and cannot be readily distinguished with X-rays. Although the KCl structure is the same as that for NaCl, the (1,1,1) reflection is now actually missing. The X-ray pattern resembles that for a cubic lattice with one half of the spacing of the actual unit cell.

We hope that these examples illustrate for you the overall conclusion that the important information in an X-ray diffraction pattern is contained in both the *positions* (values of θ) and *intensities* of the diffracted beams. Analysis of the former (i.e. values

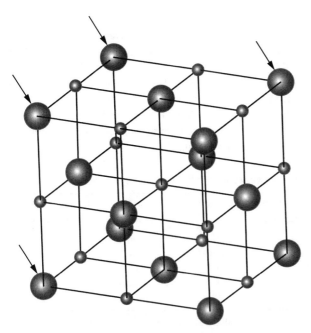

Figure 6.18 View of the sodium chloride unit cell showing the (1,1,1) planes of one type of ion (indicated) and interstitial planes comprising ions of the other type.

of θ) usually allow recognition of the unit cell and lattice type and to the dimensions of the unit cell. The intensities of the reflections depend on the *nature* of the ions present and these are the key to the application of X-ray diffraction for structure determination for molecules, as illustrated in subsequent sections.

Ionic radii

The measurement of the unit cell length can also lead to values of the ionic radii of the constituent ions. For example, for NaCl the length of the cell side is equal to the sum of the diameters of the sodium and chloride ions (see Figure 6.12; a full analysis is presented as a worked example in the next section). In LiCl, which has the same type of structure, the Li^+ ions are so much smaller than the Cl^- ions that the latter actually 'touch' along the diagonal of a face (see Figure 6.12). From the X-ray powder photograph from this compound, the length of the cell side can be obtained (0.51 nm; 5.1 Å) and used to calculate the length of the diagonal of the face and hence the ionic radius of Cl^-. Then, from the length of the cell side for NaCl, you should be able to calculate the ionic radius of Na^+ (see Exercise 6.3).

Worked examples; structural analysis of unit cells from powder diffraction patterns

Worked example 6.1. We aim to show how consideration of the powder X-ray diffraction pattern from metallic copper (shown in Figure 6.4) can lead to the determination of the unit cell type (cubic, fcc or bcc) and to the measurement of the size of the unit cell. Note that Figure 6.4 was recorded with X-rays of wavelength 0.154 nm, and the film, which is reproduced to scale, has 1° equivalent to 1mm.

Firstly, we should note that the angle between the two arcs for a given cone from the diffracted beam (i.e. where it cuts the film) can be measured directly from the figure; this angle equals 4θ, where θ is the angle of incidence of the beam on the particular plane of the crystal causing diffraction (see Figures 6.7 and 6.8); the angle between the arc and the centre of the film is thus 2θ (see Figure 6.8). So we should measure θ values from the film, converting from the distances measured in mm.

Next we should convert these to values of $\sin^2\theta$, and then look for a simple integral ratio between these values (as expected from the application of Equation (6.5) relating diffracted angles to planes in the crystal).

Table 6.4 Analysis of the X-ray powder diffraction photograph for copper

θ/degrees (taken from the photograph)*	$\sin\theta$	$\sin^2\theta$	Integral ratios $(h^2 + k^2 + l^2)$	h, k, l
22.0	0.3746	0.1403	3	1,1,1
25.5	0.4305	0.1853	4	2,0,0
37.3	0.6060	0.3672	8	2,2,0
45.0	0.7071	0.5000	11	3,1,1
47.8	0.7408	0.5488	12	2,2,2

*See Figure 6.4. Angle between line and centre of film corresponds to 2θ: 1 mm $\equiv$ 1°

You are encouraged to confirm these measurements and calculations, which are summarized in Table 6.4. As you can see, these are found to be in the ratios of 3:4:8:11, which corresponds to the solutions to Equation (6.5) for a *face-centred* cubic lattice, with reflections from the planes with h,k,l (1,1,1), (2,0,0), (2,2,0), and (3,1,1).

Any one of these the values of θ (and the associated values of h,k,l) with the given value of the wavelength can be used to give a measure of a, the length of the side of the unit cell, as 0.356 nm (3.56 Å).

Worked example 6.2. When X-rays of wavelength 0.154 nm impinge on a single crystal of NaCl the X-ray diffraction pattern observed in Figure 6.5 is produced. You are invited to analyse the pattern, noting as above that the film is reproduced to scale, with $1° \equiv 1$mm, to determine the lattice type (cubic, bcc, or fcc) and determine the length of the side of the unit cell.

Firstly, you should analyse the X-ray diffraction pattern as carried out for the single crystal of copper described above. We suggest therefore that you measure values of θ, say for the first four lines, and then tabulate values of θ, and $\sin^2\theta$ and attempt to establish the simplest integral values that relate these. Results of measurements and calculations are given in Table 6.5.

Table 6.5 Analysis of the X-ray powder diffraction photograph for sodium chloride

θ/degrees*	$\sin\theta$	$\sin^2\theta$	Integral ratios $(h^2 + k^2 + l^2)$	h, k, l
13.9	0.2402	0.0586	3	1,1,1
16.0	0.2756	0.0759	4	2,0,0
22.8	0.3875	0.1502	8	2,2,0
27.0	0.4540	0.2061	11	3,1,1

*See Figure 6.5. The angle between a line and centre of film corresponds to 2θ:1mm $\equiv$ 1°. The weak lines associated with the reflections from 1,1,1 and 3,1,1 planes are almost invisible here: see text

As you can see, the simplest ratios obtained for $(h^2 + k^2 + l^2)$ are again 3:4:8:11.

To take this analysis further, why not try to predict the angle appropriate to the next line in the series, and check to see if it is present; would you expect this to be a weak or a strong line (and why?)

The answer is that the next allowed reflection will be that associated with the (2,2,2) planes, to give $(h^2 + k^2 + l^2) = 12$. This leads to a calculated $\sin^2\theta$ value of 0.2250, with $\theta = 28.30$; inspection of Figure 6.5 confirms the expected pair of lines corresponding to this value of θ. It is a strong reflection, since rays from layers of Na^+ ions and from layers of Cl^- ions should be in phase (see Figure 6.18; the (2,2,2) planes are half as far apart as the (1,1,1) planes indicated).

6.5 Structural determination for molecules

In a crystal where the lattice points are occupied by covalent molecules (e.g. in a crystalline organic compound) the diffraction from each separate atom must be considered. The molecules themselves will be symmetrically placed with respect to each other, but atoms in the molecules will now not only be found at the corners and body- or face-centres of the cells. However, there will still be characteristic 'repeat distances' in the structure (i.e. periodic variation in electron density) which cause diffraction as discussed before for simple ions, and the essential theory is the same. The X-ray examination is carried out for a single crystal, if one is available, and a very complicated pattern of reflections, with different intensities, is obtained (an example is given later: see Figure 6.22). The problem then is to work back from this information to a plot of the electron density (which causes diffraction) in the unit cell. An electron density contour map, in which the 'peaks' correspond to atoms in the unit cell, can then be drawn. For example, Figure 6.19 shows an electron density map for benzene determined in an early classic study using X-ray diffraction, and the positions of the carbon atoms (a symmetrical hexagon, with equal C–C bond lengths) can be seen: peaks from hydrogen atoms are not clearly observed because there is relatively little X-ray scattering from these atoms (which depends on the number of electrons around a given atom).

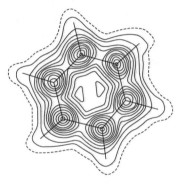

Figure 6.19 Electron density map for benzene. Reproduced from the paper by E. G. Cox, D. W. J. Cruickshank and J. A. S. Smith, *Proceedings of the Royal Society*, 1958, **247A**, p. 1.

The use of modern computers and advances in instrumentation make this type of analysis and presentation relatively straightforward (though the details are beyond the scope of this book), and it is now becoming routine to obtain the complete molecular structure of many organic and organometallic molecules, giving a detailed picture of the molecules and accurate measurements of bond lengths and angles. (Note that these are average values: see page 38.)

The following two examples illustrate the power of the X-ray diffraction approach in the study of organic and organometallic molecules, even when conventional spectroscopic methods, with mass spectrometry, have been employed to obtain the molecular weight, formula and proposed molecular structure. In the X-ray diffraction approach, the structures derived from the diffraction patterns from a single crystal are typically shown in a form in which the 3-dimensional structure of the molecule can be visualized (e.g. as in the so-called ORTEP diagram used in the first example).

Oxidation of the molybdenum complex $Mo(CO)_3(\eta^5-C_5H_5)$ in the presence of $AgBF_4$ and 4-methoxy-6-methyl-2-pyrone results in the formation of a compound

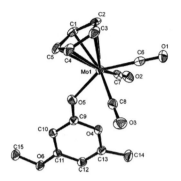

Figure 6.20 ORTEP structural diagram of a complex of molybdenum containing carbonyl, cyclopentadienyl and 2-pyrone ligands. Reproduced with permission from *Organometallics*, 2004, **23**, 4964–4969, I. J. S. Fairlamb, J. M. Lynam, I. E. Taylor and A. C. Whitwood.

whose structure can be determined, and shown in Figure 6.20. Spectroscopic results (^{1}H NMR, ^{13}C NMR, IR, with MS) led to a suggested structure which was confirmed by X-ray diffraction results. This shows clearly the central metal atom, the three carbonyl molecules attached to the metal, the cyclopentadienyl unit bonded symmetrically to each of the carbon atoms (this is referred to as η-bonding), and finally the pyrone ring η-bonded, through the oxygen attached to C9. Importantly, the diffraction pattern allows us to measure bond angles and lengths with precision—and in a way which may often reveal steric and electronic effects within a molecule. (To consider this structural result in more detail, e.g. to rotate the molecule to view it from different directions, you are encouraged to refer to the Online Resource)

X-ray diffraction is also particularly useful for determining the precise (or absolute) stereochemistry of a molecule, especially for identifying which stereoisomer (from a variety of possible isomers) is under investigation. This is relevant when the structures contain different possible geometrical isomers or optical isomers which derive from the presence of one or more stereogenic centres. An example is shown in Figure 6.21 which also shows an alternative way to depict and visualize the 3D structure. Note the nitrogen atom and the methyl, ethyl and phenyl groups, as well as the stereochemistry at each of the three chiral centres; this compound is correctly described as the (*S,S,S*) stereoisomer of the compound. Again, you are encouraged to visit the on-line resource.

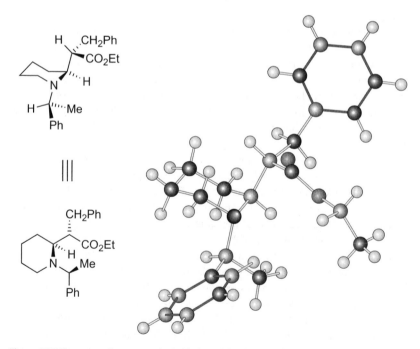

Figure 6.21 X-ray crystal structure of ethyl (2*S*)-3-phenyl{(2*S*)-1-[(1*S*)-1-phenylethyl]piperidine-2-yl}propionate. Reproduced with permission from *Org. Biomol. Chem.*, 2007, **5**, 3614–3622; J-P. R. Hermet, A. Viterisi, J. M. Wright, M. J. McGrath, P. O'Brien, A. C. Whitwood and J. Gilday.

X-ray studies of biological molecules

X-ray diffraction also provides an excellent method for investigating the structures of biologically important molecules which contain repeating chemical groups (e.g. proteins and nucleic acids). One particularly important and relatively early example of particular historic and scientific importance was the analysis of the X-ray pattern from the nucleic acid DNA. The X-ray measurements of Rosalind Franklin and their interpretation by James Watson and Francis Crick were described in classic research papers to the journal *Nature* in 1953. The diffraction patterns are interpretable in terms of this molecule having an interwoven double-helix of repeating units of base pairs. X-ray reflections indicate characteristic 'repeat distances' (*cf.* planes in a crystal) of 0.34, 3.4, and 2.0 nm (3.4, 34 and 20 Å): these are, respectively, the distance between successive units in the chain, the repeat distance (pitch of the helix), and width of the spiral. This new information enabled the essential features of the DNA structure to be elucidated by Watson and Crick in their pioneering and Nobel Prize-winning research.

Computer graphics now enable electron-density and molecular structure maps to be readily visualized to give a three-dimensional view of complex structures, as noted earlier, and for these to be 'rotated' to reveal full details of the structure. For example, reference to the website given in the online resource allows you to see the full detail of the structure of DNA—its phosphate groups, sugars, and purine and pyrimidine bases, showing base-pairing, helical chains, etc: you are encouraged to study this and manipulate the structure in 3D to explore aspects of the structure of this remarkable molecule.

Another example (Figure 6.22) shows the X-ray diffraction pattern from a crystal of haemoglobin and a side view of part of the molecule, derived from the electron-density map. The haem ring, containing the large Fe atom with an oxygen molecule bonded above it, can clearly be seen. Those amino acids in the protein chain which are closest to the haem are shown; the histidine residues situated above and below the metal can be readily identified. This type of information, nowadays greatly assisted by employing X-rays from synchrotrons and advances in computing technologies, is vitally important to the understanding of the structural and biological functions of highly complex molecules such as proteins and enzymes, and also the subsequent design of medicines (referred to as structure-based drug discovery).

These types of application owe much to the dedicated and pioneering research of Dorothy Hodgkin. Much of this was carried out before the advent of computer analysis and involved years of complicated analysis of the complicated diffraction patterns of spots of varying intensity. Particular highlights include her complete structural analysis in the early 1940s of a derivative of the antibiotic penicillin, which revealed the presence of the novel β-lactam feature (a 4-membered ring rather than the alternative thiazolidine-oxazolone ring) (refer to the cover of this book for the original hand-drawn electron density map and 3D-structure). She also successfully revealed the structure of vitamin B_{12} (for which she was awarded the Nobel Prize in 1969, after 34 years' investigation) and insulin (the protein hormone which contains 51 amino acids) and for which the sequence of amino acids had been established by Frederick Sanger.

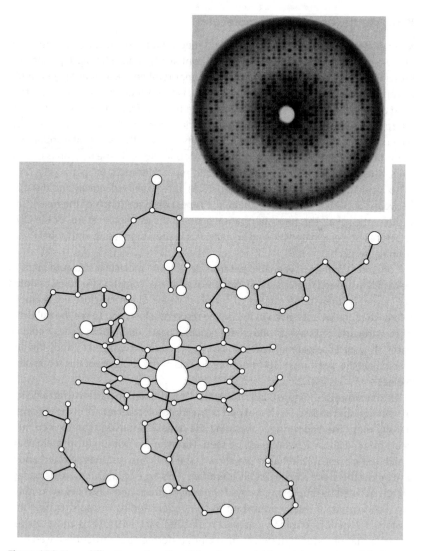

Figure 6.22 X-ray diffraction pattern (inset) from haemoglobin and the structure of the haem ring, iron atom, and associated amino acids.

In reflecting on the use of structural analysis (X-ray diffraction in particular), how better than to conclude with words of Max Perutz, himself a Nobel Prize winner, and quoted in *Max Perutz and the Secret of Life*, Georgia Ferry, London, Chatto and Windus, 2008?

'Why water boils at 100°C and methane at –161°C, why blood is red and grass is green, why diamond is hard and wax is soft, why graphite writes on paper and silk is strong, why glaciers flow and iron gets hard when you hammer it, how muscles contract, how sunlight makes plants grow and how living organisms have been able to evolve into ever more complex forms ... The answers to all these problems have come from structural analysis'.[2]

[2] An excellent short survey of the revolution in biological science brought about by the use of X-rays over the last 100 years is given by Clare Sansom, *Chemistry World*, 26 August 2014

6.6 Neutron diffraction

One of the disadvantages of X-ray diffraction is that atoms of low atomic number are difficult to locate in the presence of heavier elements (in particular, hydrogen atoms are difficult to detect even in the presence of carbon or oxygen) because the intensity of scatter (or 'reflection') from an atom depends on the number of electrons around that atom.

If a study of the precise positions of hydrogen atoms is to be made (e.g. a study of a metal hydride or a hydrated salt) then the related method of **neutron diffraction** can be employed. In this technique a beam of neutrons (from an atomic pile) is diffracted in exactly the same manner as is a beam of X-rays, and the intensity of the beam scattered in different directions is monitored. There is much less variation with atomic number in the scattering of neutrons than there is for X-rays, so that the technique becomes a more sensitive probe for hydrogen atoms than is X-ray diffraction. Obviously neutron diffraction is both difficult and comparatively expensive, and it is only usually attempted when a full X-ray analysis (to locate all the heavier atoms) has been performed.

6.7 Electron diffraction—method and structure determination

At approximately the same time that X-ray diffraction was first demonstrated, other investigators showed that electrons possess wave-character and that they too can exhibit interference patterns.

For example, it was shown that when an electron beam is fired through a very thin gold foil, a diffraction pattern consisting of concentric dark rings appears on a photographic plate behind the metal. This type of experiment does not find wide application, because electrons tend to be absorbed by, rather than scattered by, denser forms of matter (liquids, solids). For this reason most of the electron diffraction applications in the field of molecular structure have been concerned with gaseous samples.

Theory and instrumentation

The sample under investigation is usually examined at low pressure (*ca.* 10^{-4} N m^{-2}) and the incoming electron beam is scattered by the electric field of the atoms, provided by the nuclei and electrons.

A single atom scatters the electron beam in all directions; however, for a molecule the rays scattered from the constituent atoms (see, e.g. Figure 6.23) will be in phase only in certain directions. These will depend on λ (for the electrons) and the path difference, $d(AB)$, which itself depends on the bond length and the angle made by the bond to the direction of the beam (*cf.* the Bragg Equation). For a large collection of such molecules there will be many molecules at the appropriate angle θ for in-phase reflection, so that a beam which consists of a cone of diffracted electrons is produced (as with X-ray diffraction from powders). There will be more than one angle at which this occurs, and the result is recorded on a photographic film as a series of dark and light rings (see Figure 6.24); the intensity variation is usually expressed as

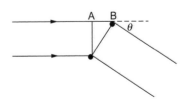

Figure 6.23 Diffraction of an electron beam by a diatomic molecule.

an experimental distribution curve or radial distribution function (showing how the observed intensity varies in a line going out from the centre) which can be studied directly and also compared with a computed function (modelled on a proposed structure).

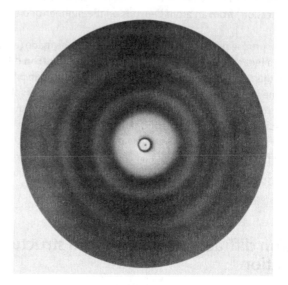

Figure 6.24 Electron diffraction pattern from tetrafluoroethene.

The beam of electrons used in the experiments must have a wavelength somewhat smaller than typical bond lengths if diffraction is to be detected (as with X-rays). This is arranged as follows. The wavelength of an electron depends on the mass and the velocity of the electron m_e and v (the **de Broglie** Equation (6.5)).

$$\lambda = h/m_e v \tag{6.5}$$

The velocity of the electrons can be altered using an accelerating potential V (see also page 103, section 5.2), so that

$$\tfrac{1}{2}m_e v^2 = eV \tag{6.6}$$

Combining these two equations it follows that

$$\lambda = h\sqrt{\frac{1}{2m_e eV}} \tag{6.7}$$

With $V = 40$ kV, λ is *ca.* 0.006 nm, and this proves suitable for electron diffraction studies.

Structure Determination

Gas-phase electron diffraction is normally carried out only for small molecules. The necessity for this simplification arises because the separations (see Figure 6.23) which each give rise to diffracted 'cones' are those between bonded *and* between non-bonded pairs of atoms in a molecule. These cones are recognized as peaks in the radial distribution function: the number of peaks corresponds to the number of such separations, and the analysis using the Bragg Equation gives appropriate distances.

A particularly simple pattern arises from the molecule P_4 (which is concluded to be a regular tetrahedron) which has only a single (P···· P) distance causing diffraction. Diatomic molecules such as iodine (I_2) also give a very simple pattern and analysis from which the average bond length of 0.267 nm (2.67 Å) is calculated.

For CO_2, the radial distribution curve indicates two internuclear distances (C–O, O–O), one twice the other, indicating a linear molecule with r_{CO} 0.115 nm, 1.15 Å. (*cf.* pages 10, 11, section 2.4; 15, section 2.5; 16, 17, section 2.6; 20, 21, section 2.8; and 34–37, section 2.13).

For the molecule CCl_4 for example, there will be scattering at certain angles from the C–Cl bonds but also from the Cl....Cl pairs in the same molecule (i.e. these pairs also behave as a diffraction grating, as with the oxygen atoms in CO_2). From the number and spacing of the rings in this fairly simple example it is possible to deduce that the molecule is *tetrahedral* (rather than square planar, and that the single average C–Cl bond length is 0.177 nm (1.77 Å).

If the molecule under investigation is a little more complicated then there are more separations to confuse the analysis. For CF_3Cl, for instance, the distances responsible for diffraction are C–Cl, C–F, F....Cl and F....F. A detailed solution must also allow for the different types of atom (different scattering intensities), and in practice it can be carried out to give bond lengths to about 0.001 nm and to confirm the molecular shape. In this respect the results are similar to those obtained using infrared spectroscopy (see sections 2.4-2.6, 2.8, 2.13).

The method also can be extended to more complicated molecules (though electron diffraction is rarely, if ever, used as a service instrument for routine structure determination). For example, the electron diffraction pattern for benzene provides evidence for a molecule with three separate interatomic C....C distances and four interatomic C....H distances, consistent with a regular planar hexagonal arrangement of the six carbon atoms in the molecule (all C–C bonds the same length, 0.139 nm, 1.39 Å). In contrast, the cyclooctatetraene molecule (C_8H_8), can be shown to be tub-shaped, with alternating long and short bonds as shown in Figure 6.25. Study of the molecule 1,2-dichloroethane reveals the predominance of a *trans*-conformation with the two carbon-chlorine bonds making an angle of 180° to each other (i.e. to minimize steric interactions involving the two large chlorine atoms).

Other notable gas-phase examples include studies of cyclohexane, by the Nobel Prize winner Odd Hassel and his colleagues, which demonstrate the preferred chair-like shape (*conformation*) of the six-membered ring, as shown in Figure 6.26, and also those which confirmed the structure of Buckminsterfullerene C_{60} (Figure 6.27) whose existence had been revealed through mass spectrometry studies (X-ray studies of the crystalline material had been thwarted because the molecules readily rotate, even in the solid state). The cyclohexane study was particularly notable both for determining the C–C and C–H bond lengths, but also for establishing clearly the two different types of C–H bond (now referred to as axial and equatorial). (See O. Hassell and H. Viervoll, *Acta Chem. Scand.* 1947, p. 147.)

Figure 6.25 Structure of cyclooctatetraene.

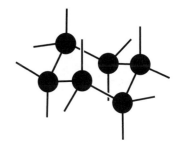

Figure 6.26

Figure 6.27

Other recent applications include the use of time-resolved methods involving pulses and electrons, produced by ultra-fast (femtosecond) pulsed laser irradiation of a gold photocathode, which can be used to study the structure of short-lived species generated in photo-induced gas-phase reactions. And beams of electrons can be used in diffraction experiments with solids, as with X-rays: these have wavelengths smaller than those used in typical X-ray experiments, and can be employed, in transmittance or reflectance mode with thin films and very small crystals (e.g. of novel inorganic materials used in magnetic devices).

Low energy electron diffraction

This method (LEED) can be employed to probe the arrangement of atoms in a solid sur- face; electrons with a wavelength in the range 0.5–0.05 nm (and hence with energy lower than that associated with gas-phase electron diffraction) are directed on to the sample (e.g. a metal). Electrons are reflected from the top layer of atoms, and the diffracted in-phase beam in certain directions (*cf.* the Bragg Equation) is detected on a fluorescent screen. The resulting pattern leads to details of the two-dimensional structure of the top layer.

The method can also be applied to the study of the adsorption of gases on solids, and in this way it has provided a useful technique for recent research on the mechan- isms by which some solids act as catalysts.

6.8 Summary

As a result of studying this chapter you should be able to appreciate and understand:

- how X-rays are produced and diffracted by the regular inter-ionic and inter- atomic spacings in a lattice;

- how X-ray diffraction patterns from ionic lattices and single crystals can reveal structural information via analysis based on the Bragg Equation; this will include determination of cell type, Bravais lattice (e.g. *bcc* vs. *fcc*), with measurement of cell dimensions;

- how simple lattices and planes can be described and related to X-ray diffraction patterns, via the use of Miller Indices;

- how, in outline, electron-density maps and 3D-structures can be derived from X-ray diffraction patterns, so that full molecular structure details can be obtained, including bond lengths and angles (hence complementing and expanding the conclusions on molecular structure of unknown molecules obtained by use of the techniques described earlier in this book). You should, in particular, be able to appreciate the important role of X-ray diffraction in revealing the structure of highly complex molecules, including those of especial biological importance, like DNA and haemoglobin;

You should also have developed skills in the analysis of some relatively simple pat- terns, via worked examples and problem-solving.

Finally, you should have gained an appreciation of the basic principles and uses of Neutron and Electron Diffraction, with emphasis on the different types of application of these methods compared with the use of X-rays as described in much greater detail earlier.

6.9 Exercises: Analysis of the X-ray diffraction patterns of crystals

Worked solutions to the exercises are available on the Online Resource

6.1 Figure 6.3 shows the X-ray powder photograph from molybdenum (recorded with X-rays of wavelength 0.154 nm); the film has $1° = 1mm$ and is reproduced to scale. From the positions of the (2,0,0) reflections calculate:

(a) the Bragg angle for this reflection;

(b) the length of the side of the unit cell for molybdenum.

6.2 Figure 6.6 is the single crystal photograph (to scale) obtained using X-rays of wavelength 0.154 nm from a crystal of sodium chloride with one axis vertical. Figure 6.8 and Equation (6.2) describe how the layer lines arise. The angle φ for any particular layer line can be obtained as follows:

$$\tan\varphi = y / R$$

where R is the radius of the film (29 mm in this example) and y is the vertical distance between the layer line and the $n = 0$ layer.

(a) Calculate φ for the $n = 1$ layer and hence calculate the value of a, the length of side of the unit cell.

(b) Why are the spots in the $n = 1$ and $n = -1$ layers weaker than those in the $n = 2$ and $n = -2$ layers?

6.3 Lithium chloride has a sodium chloride-type lattice, with the length of the side of the unit cell 0.513 nm. On the assumption that the chloride ions actually 'touch' along the diagonal of the face of the unit cell, derive a value for the ionic radius of Cl^-. Use this value together with the length of the cell side for NaCl (0.56 nm) to derive the ionic radius of Na^+.

6.4 The powder photograph of LiCl has reflections at increasing angles (θ) which have values of $\sin^2\theta$ in the ratio 3, 4, 8, 11, 12, etc. (like NaCl). Unlike NaCl, for which the reflections associated with some of these (3 and 11) are much weaker than the others, all the reflections for LiCl have approximately the same intensity. Can you account for this observation?

6.5 In a gas-phase electron diffraction study of AsI_3, peaks in the radial distribution curve correspond to *two* internuclear separations of 0.254 nm (attributed to the As–I bond) and 0.385 nm (the I⋯I separation). What can you conclude about the geometry of AsI_3?

6.6 In a gas phase electron diffraction study of 1-chlorocyclohexane, it was found that the chorine atom occupied solely a conformation with the chlorine atom replacing a hydrogen atom in a single position (axial or equatorial). Predict, giving your reason(s) whether this is the axial or equatorial C–H.

6.10 Further Reading

W. Clegg (2015), *X-Ray Crystallography*, 2nd edition, Oxford Chemistry Primers, Oxford University Press, Oxford.

Glossary

Absolute stereochemistry A full 3-dimensional structure of a molecule, in which, for example, the R- and S- or D- and L- forms of optically active molecules are distinguishable (as with left or right hands).

Absorbance (A) Absorbance is defined as the logarithm of the ratio of the intensity of the incident radiation (I_0) to that of the transmitted radiation (I) in a UV–visible absorption experiment.

Activation enthalpy ($\Delta H^{\ddagger}$) The term used to describe the energy barrier in a reaction at constant pressure. The larger the barrier the more dependent the rate constant will be on temperature.

Anti-bonding orbitals An anti-bonding orbital is a molecular orbital whose occupation by electrons actively weakens the overall bond strength of a molecule.

Anti-Stokes scattering Inelastic scattering of light in a Raman experiment in which the light takes up energy from a molecule, leaving the latter in a lower quantum state.

API Atmospheric pressure ionization mass spectrometer.

Arrhenius plot A plot of ln$k$ (rate constant) vs. $1/T$ in reaction kinetic studies is known as an Arrhenius plot. The slope, $-E_a/RT$, yields the activation energy, E_a.

Asymmetric top An asymmetric top has all three moments of inertia unequal.

Backward wave oscillator A development of the klystron, used as a source of microwave radiation in rotational spectroscopy.

Base peak The most intense peak in a mass spectrum.

Beer–Lambert Law The dependence of light absorbance (A) on the concentration of a solution, c, and the path length, d, in the solution is given by the Beer–Lambert Law, $A = \varepsilon cd$, where ε is known as the molar decadic absorption coefficient.

Bragg Equation The constructive interference of X-rays diffracting from lattice planes in a crystal occurs when the path-length difference between rays diffracting from different planes is equal to integer (n) multiples of the wavelength, λ. The path length depends both on the distance between lattice planes, d, and the glancing angle of the radiation, θ. This relationship is given by Bragg's Equation, $n\lambda = 2d\sin\theta$.

Bravais lattice Nomenclature used to describe different types of lattices within a given unit cell type (with the same overall symmetry); for example, for cubic systems, a *simple* (or *primitive*), *face-centered*, and *body-centered* cubic lattice.

CCD Charge-coupled detector, a device used to detect and quantify the intensity of incident radiation, e.g. in X-ray diffraction experiments.

Chemical shift In an NMR experiment, the chemical shift of a nucleus is the difference between its resonance field or frequency and that of a known standard. For protons, that standard is tetramethylsilane, commonly referred to as TMS. The shift arises because the applied magnetic field induces a circulation of the electrons in the molecule and this circulation creates additional small magnetic fields. The strength and sign of these induced fields depend on the local electronic structure nearby the magnetic nucleus of interest.

Chirped pulse A microwave pulse in which the frequency at the end of the pulse is much higher than the frequency at the beginning of the pulse. Used in Fourier Transform microwave spectroscopy.

Chromophores Groups with characteristic optical absorptions are known as chromophores. Typical examples might include the C=O and C=C groups. The first of these absorbs at about 280 nm as a result of an n-to-π^* transition (where n denotes a non-bonding orbital) whilst the second absorbs at about 180 nm as a result of a π-to-π^* transition.

CI Chemical ionization source in a mass spectrometer.

Conformations The different arrangements of atoms in a molecule which result from rotation around a single bond (e.g. C–C) are called conformations.

Conjugation Conjugation occurs in molecules with chains of alternating single and double carbon–carbon bonds (although may involve other atoms such as oxygen and nitrogen). In such systems, the π-bonding is delocalized over as many of the sp^2-hybridized carbon atoms as are present in the chain. In ethene for example, the π-bonding is *localized* over two carbon atoms, whilst in buta-1,3-diene it is *delocalized* over four carbon atoms.

Constructive interference The interaction between two in-phase waves (such as encountered in X-ray diffraction or in the overlap of atomic orbitals) results in constructive interference in which the intensity of the resultant wave is the sum of the component intensities.

Continuous wave NMR Method for detecting NMR spectra in which the incident frequency is held constant and the magnetic field is varied until the condition for resonance is met.

COSY Correlation spectroscopy, a term used to describe the presentation of two-dimensional NMR spectra in experiments in which multiple pulses are employed and in which coupling partners are revealed.

Cubic lattice A structure with a cubic unit cell, having all cell sides the same length ($a = b = c$) and all internal angles 90°.

Decoupling This refers to experiments in which the splittings in a given NMR signal can be removed by irradiating the second nucleus so that it undergoes rapid transitions rather than providing a static local magnetic field for the neighbouring nucleus to interact with.

Degeneracy If two or more states of a quantum mechanical system yield exactly the same measurable energy, then they are said to be degenerate.

DEPT The acronym for Distortionless Enhancement of Polarization Transfer, a means whereby the number of protons on a given carbon atom can be distinguished in ^{13}C NMR experiments.

Deshielding The situation in NMR spectra when the local field at a given nucleus (compared to the applied field) is increased *or* the shielding effect is decreased (see shielding).

Destructive interference The interaction between two out-of-phase waves results in destructive interference in which the component wave intensities cancel each other out.

Diode array A linear array of photodiodes used to monitor the intensity of u.v./visible radiation emerging from a dispersion device. The intensity of the light at particular points along the array provides a means of relating the light intensity to wavelength.

Dipole moment (molecular) The molecular dipole moment is an electric dipole resulting from the unequal distribution of charge (see electric dipole).

Dispersion The effect of a prism or grating in causing different colours of light to refract at different angles. Dispersion of polychromatic light or white light into its component wavelengths produces a rainbow.

Distribution curve The variation of intensity with distance from the incident beam centre in an electron diffraction experiment.

Elastic scattering Scattering of light in which no energy is exchanged (see Rayleigh scattering).

Electric dipole A measure of the separation of positive and negative electrical charges in a system of electric charges.

Electromagnetic spectrum The range of all possible wavelengths of electromagnetic radiation extending from radiowaves at the very long wavelength end to gamma waves at the shortest wavelength end.

Electron impact (EI) Method for producing positive ions (for mass spectrometry) via high-energy electron impact collisions.

Electrostatic analyser Method for further separating ions of similar m/z in a mass spectrometer.

Energy barrier See activation enthalpy.

ESI Electrospray ionization as source of ions in a mass spectrometer.

Fermi resonance A resonance involving the fundamental level of one mode interacting with an overtone in another mode as a consequence of an accidental near-degeneracy. The result is a doublet in an infrared or Raman spectrum with the two lines shifted slightly from their native positions.

Field-free region The region in the flight path of the ions in a mass spectrometer in which there are no electric or magnetic fields. Sometimes referred to as the drift region in which the ions are no longer subject to acceleration or deceleration and travel at constant velocities.

Force constant The force constant, k, provides a measure of the stiffness of a chemical bond.

FT, Fourier Transform A mathematical transformation used to transform signals from the time domain into the frequency domain. In NMR spectroscopy, the time-domain free induction decay that follows the radiofrequency pulse is Fourier-transformed into a Lorentzian line-shape in the frequency domain. The result is a spectrum of lines presented in frequency terms. A similar approach is used in microwave spectroscopy.

Fundamental absorption A vibrational transition in which the molecule is lifted from its zero point level into the first excited vibrational level, corresponding to $v = 1$. In an infrared spectrum, the $\Delta v = +1$ transition usually appears with much stronger intensity than overtone transitions, $\Delta v = +2, +3, \ldots$ symmetry considerations otherwise allowing.

GC/MS Gas chromatography/mass spectrometry, for sample separation and mass analysis.

GPC/MS Gel permeation chromatography/mass spectrometry, for sample separation and mass analysis.

Group vibration A vibration which localizes in a particular part of the molecule, often associated with the presence of a particular functional group.

Halogen bond A weak interaction involving a halogen atom as an acceptor of electron density.

Hydrogen bond An electrostatic interaction between polar molecules in which a hydrogen atom bound to an electronegative atom experiences an attraction to another electronegative atom, thereby acting as an acceptor of electron density.

Indexing A term used in the analysis of powder diffraction photographs, in which reflections are analysed in terms of possible values of the Miller Indices h, k, and l, leading to identification of the cell type and Bravais lattice (and also the cell dimensions).

Inelastic scattering Scattering of light in which energy is exchanged between the photon and the collision partner, typically a molecule (see Stokes and Anti-Stokes scattering).

Infrared active A vibrational mode is infrared active if a changing dipole results from the vibration. The fundamentals of such modes may then appear in an infrared spectrum.

Infrared light Electromagnetic radiation with wavelengths between 770 nm and about 1000 μm.

Integration trace A plot which gives 'steps' or increments in an NMR spectrum, which are proportional to the relative intensities (abundances) of the atoms which give rise to the absorptions.

Ionic lattice Lattices which contain two or more ions (e.g. Na^+ and Cl^-) at the lattice points.

Ionization (source) Method for achieving ionization, typically through loss of an electron to form a cation.

J The rotational quantum number, taking values 0, 1, 2, 3, ...

J_{HH} The spin-spin splitting (coupling) interaction between non-equivalent protons (usually neighbours) in an ^1H NMR experiment.

J_{max} The rotational quantum number corresponding to the rotational level having the highest population at a given temperature.

Klystron A device used as a source of microwave radiation.

Lattice The term used to describe the regular array of ions, atoms or covalent molecules in a crystal.

Magnetic focusing device The use of an applied magnetic field to focus ions of a given m/z in a mass spectrometer.

MALDI Matrix Assisted Laser-Desorption Ionization for vaporization of substrates and generation of positive ions in a mass spectrometer.

Mass analyser The section of a mass spectrometer that separates ions on the basis of their m/z values.

Mass resolution The precision with which ions of differing masses can be distinguished in a mass spectrometer.

Mass spectrum Plot of the intensities of positive ions of different masses against their m/z values.

Mass-to-charge ratio, m/z The mass of an ion divided by its charge; where the ion is singly charged the m/z value will be equal to the mass of the ion (expressed as its relative molar mass).

McLafferty Rearrangement Formation of positive ions in a mass spectrometer by β-cleavage rearrangement of molecules containing a carbonyl-group.

M$^+$ peak Peak from the molecular ion in a mass spectrum.

MH$^+$ peak Peak from a protonated molecular ion in a mass spectrometer.

M+1 peak Peak from the molecular ion in a mass spectrum with a mass one unit more than the normal isotopic composition (reflecting the presence of, say, a ^{13}C atom).

Microwave radiation Electromagnetic radiation with wavelengths between 10 mm and 30 cm.

Miller Indices (h,k,l) Nomenclature used to represent different planes in a crystal. These allow interstitial distances, d, to be expressed in terms of the values of h, k and l which describe that plane.

Millimetre wave radiation Electromagnetic radiation with wavelengths between 1 mm and 10 mm.

Molar decadic absorption coefficient Sometimes referred to as the molar extinction coefficient, with symbol, ε (see Beer–Lambert Law). It provides a measure of how strongly a particular chemical species absorbs light at a given wavelength.

Molecular ion peak (M) Peak in the mass spectrum which corresponds to the intact molecular ion, prior to fragmentation in a mass spectrometer.

Moment of inertia Provides a measure of the torque or turning force required to induce rotational motion in a rigid body about a particular axis of rotation.

Monochromatic radiation Radiation with a single wavelength (within some defined bandwidth).

MRI Magnetic Resonance Imaging—a method for measuring an NMR signal e.g. from ^{1}H, using pulsed magnetic fields across a macroscopic sample, most commonly in medical applications. The method acquires a series of 2D image slices across a sample. These can then be combined to construct a 3D image, for example, of the brain of a patient.

MRS Magnetic Resonance Spectroscopy—a development of MRI in which a high resolution NMR spectrum can be recorded to monitor components at a given point in a sample (e.g. the human brain).

Mutual exclusion principle For molecules possessing a centre of inversion, all infrared active modes will be Raman inactive and all Raman modes will be infrared inactive.

n→π* transitions An electronic transition from a non-bonding orbital to an antibonding π* orbital.

π→π* transitions An electronic transition from a bonding π orbital, commonly in a conjugated molecule, to an antibonding π* orbital.

Normal modes The natural resonant vibrations of molecules. For non-linear molecules with N nuclei, there are 3N−6 normal modes; for linear molecules, there are 3N−5 normal modes.

Nuclear magnetic moment The magnetic moment which results from the spin of the nucleus for those nuclei with odd numbers of neutrons or an odd number of protons or an odd number of both.

Nuclear Overhauser Effect NOE refers to observable effects (e.g. on peak heights) in NMR spectra resulting from *through space* magnetic interactions between protons (for example) rather than *through bonds*.

Optical density See Absorbance

Overtone transitions Vibrational transition in which the molecule is lifted from its zero point level into excited vibrational levels with quantum numbers, $v > 1$. In an infrared spectrum, overtones ($\Delta v = +2, +3$ and so on) usually appear with much weaker intensity than the fundamental, symmetry considerations otherwise allowing.

P-branch In a vibration–rotation band, the lines corresponding to $\Delta J = -1$ transitions appear in the P-branch.

Pulsed NMR Experiments in which a short pulse of the appropriate radiofrequency radiation is incident upon the nucleus (or nuclei) under investigation, which then re-radiates the appropriate resonance frequency (or frequencies) for Fourier Transform analysis to yield NMR spectra.

Quadrupole analyser A method used to separate ions according to m/z in a mass spectrometer. The analyser consists of four cylindrical rods set parallel to one another, with radio frequency voltages applied across pairs of rods to generate a quadrupole. The ions are separated according to their trajectories as they travel through the oscillating electric fields applied across the rods.

Quantum mechanics Also referred to as quantum theory or quantum physics, it provides the mathematical framework for dealing with the concept of wave–particle duality that becomes apparent in nanoscale systems such as atoms and molecules. It is fundamental to an understanding of all forms of spectroscopy involving interactions of light with quantum states of matter.

Radial distribution curve See distribution curve.

Radiowaves Electromagnetic radiation with wavelengths between 30 cm and 10 m.

Raman active A vibrational mode is Raman active if the polarizability changes during the course of the vibration. The fundamentals of such modes may then appear in a Raman spectrum.

Raman scattering The inelastic scattering of light in which a collision between a photon of light and a molecule leaves the photon with a little less energy and the molecule with a little more (or *vice versa*).

Raman spectroscopy A technique employing inelastic scattering of (typically) visible light to excite transitions between rotational and/or vibrational energy levels.

Rayleigh scattering Elastically scattered light in a Raman experiment, having neither lost nor gained energy from the process.

R-branch In a vibration-rotation band, the lines corresponding to $\Delta J = +1$ transitions appear in the R-branch.

Reduced mass The effective inertia mass in a two-body system. Can be thought of as a measure of the mass being swung about during vibrational or rotational motion.

Relative atomic and molecular mass Atomic and molecular masses are commonly defined relative to the mass of the isotope ^{12}C which by definition is equal to 12. The values are dimensionless but frequently given the unit Dalton or atomic mass unit. Commonly referred to as the atomic or molecular weight.

Relaxation The process whereby nuclei in an excited state return to the ground state with loss of energy.

Residual mass Term used to refer to the molar mass of the key fragment from an amino acid [–HNCHRC(O)–] utilized in peptide decomposition analysis via mass spectrometry.

Ring current Motion of the electrons (e.g. in a benzene ring) in the presence of an applied field, which leads to the production of a local magnetic field detectable via the characteristic NMR chemical shifts of aromatic protons.

Rotational constant A spectroscopic constant inversely proportional to the moment of inertia. In a diatomic or linear polyatomic molecule, the rotational constant is given the symbol B.

Rotational spectroscopy Typically employing absorption or emission of millimetre, microwave or far infrared radiation to excite transitions between rotational energy levels. Rotational Raman spectroscopy employs inelastic scattering of visible light.

Sector instrument The use of a magnetic field to separate ions of different m/z in a mass spectrometer.

Sequencing Determination, for example of the order (sequence) of monomers in a polymer (e.g. amino acids in a protein or peptide).

Shielding The effect of a local magnetic field (from induced electron circulation) which opposes the applied magnetic field.

Spectral editing This refers to methods (e.g. DEPT) in which decoupling techniques can be used to provide simplified spectra and/ or extra information on the origin of couplings in NMR.

Spectral resolution The precision with which differences in quantum states can be discerned in spectroscopy. In optical spectroscopy, high resolution is achieved by using monochromator gratings with a very high grating density or optical sources with a very narrow bandwidth. In NMR, high resolution is achieved by ensuring as stable a magnetic field as possible. A high resolution NMR experiment will reveal spin–spin splittings.

Spherical top A spherical top has all three moments of inertia equal.

Spin–spin coupling The magnetic interaction between nuclear spins in non-equivalent nuclei resulting in splitting of NMR peaks into multiplets.

Stokes scattering Inelastic scattering of light in a Raman experiment in which the light gives up some of its energy to a molecule, leaving the latter in a higher quantum state.

Supersonic free jet Expansion of a gas into a vacuum through a small orifice. If the orifice is considerably larger than the mean free path of the gas molecules, collisions in the high-density region near the orifice will result in a dense jet with a narrow velocity distribution that is kinetically cold.

Symmetric top A symmetric top molecule has two of its principal moments of inertia equal with the third non-zero. Depending on the relative magnitude of the unequal moment, may be either prolate or oblate.

Systematic absence The absence of X-ray reflections associated with a given plane in a crystal as a result of out-of-phase reflections from other planes situated half-way between the main crystal planes (e.g. the plane containing the ions at the body-centres of the unit cells).

Tandem, MS/MS The use of two sequential mass analysers in a mass spectrometer in which ions of a given m/z separated in the first are then allowed to decompose further in the second.

Term values Energy level expressions expressed in units of wavenumber.

TOF The time-of-flight method for mass analysis. Ions of different mass will acquire different kinetic energies when subject to an electric field gradient and will then take different times to reach a detector located at the end of a flight tube.

Transmittance The fraction of incident light transmitted through a sample at a given wavelength in an optical absorption experiment. Defined as I/I_0, where I_0 is the intensity of the incident light and I, the intensity of the transmitted light.

Two-dimensional (2D) NMR A range of NMR techniques (e.g. COSY, NOESY) in which the length of time between NMR pulses is varied, thereby adding an additional dimension to the spectra. Such methods provide much better resolution of signals which would normally overlap in a 1D NMR spectrum.

Ultraviolet radiation Electromagnetic radiation with wavelengths between 200 and 390 nm.

Unit cell The repeating building block in a three-dimensional crystal structure.

ν The vibrational quantum number, taking values 0, 1, 2, 3, ….

Vibrational spectroscopy Typically employing absorption or emission of infrared radiation to excite transitions between vibrational energy levels. Vibrational Raman spectroscopy employs inelastic scattering of visible light.

Vibration wavenumber The vibration wavenumber, ω_e, is a spectroscopic constant approximately equal to the spacing between the zero point level and the first vibrationally excited level. It is commonly, although incorrectly, referred to as the vibrational frequency.

Visible light Electromagnetic radiation with wavelengths between 390 and 770 nm.

Wavenumber, $\tilde{\nu}$ The spectroscopic unit, cm^{-1}, favoured when working in particular parts of the electromagnetic spectrum.

Sometimes referred to as the reciprocal centimetre, having dimensions of reciprocal length.

X-rays X-radiation consists of light with wavelengths between 0.01 and 10 nm.

Zero band gap The gap in the centre of a vibrational–rotation band between the first member of the R branch, $R(0)$ and the first member of the P branch, $P(1)$.

Zero point energy According to quantum theory, a vibrating molecule can never be at rest and the nuclei never completely stationary. Even at absolute zero, all molecules possess a small amount of vibrational energy known as the zero point energy.

Index

Absolute
  stereochemistry 144
Absorbance 54, 56
ACE 124
ACE inhibitors 124
Acetophenone
  MS 115
  UV 57, 60, 61
Acetylacetone
  NMR 86
  UV 63
Activation energy 88
Activation enthalpy 88
Aldehydes, see alkanals
Alkanals
  IR 42, 43
  NMR 71, 72, 75, 77, 93
  UV 59
Alkanes
  IR 40, 42, 44
  microwave 17
  NMR 75, 93, 101
  UV 58
Alkanoate esters 59, 93
Alkanoic acids
  IR 42, 45
  MS 107, 127
  NMR 75, 93
Alkanols
  IR 42, 43
  MS 107, 109
  NMR 71, 75, 86, 93
  UV 58
Alkanoyl anhydrides
  IR 43
  NMR 93
Alkanoyl chlorides, IR 42, 43
Alkenes
  IR 39, 40, 42, 44, 45
  MS 105
  NMR 75, 83, 93, 96
  UV 57–60, 62
Alkyl chains
  IR 39, 42, 44
  NMR 76–85, 94–97
Alkynes
  IR 39, 42
  NMR 76, 93
Amides
  IR 42, 43, 46, 47
  MS 109, 121, 122, 125, 128
  NMR 93
Amines
  IR 42, 46
  MS 109, 113, 114

NMR 75, 93
Amino acids
  Sequences in peptides 121–126
  Table of 122
Angiotensin II
  MS 124
  NMR 87
Angular momentum 3
Anharmonic
  oscillator 25–27, 35
  potential 25, 38
Anharmonicity constants 25
Antibody 66
Antibonding orbitals 56–61
Antigen 66
API, APCI in MS 105, 107
Anti-Stokes scattering 19–22, 28, 29
Arenes
  IR 42, 43, 45
  MS 111
  NMR 73, 75, 76, 82, 84, 93
  UV 55, 60
  See also Benzene
Arrhenius plot 88
Arsenic triiodide 151
Aston, Francis 102
Asymmetric top
  oblate near-symmetric 16
  prolate near-symmetric 16
Atomic orbitals 53, 58
Atomic spectra 3
Atomic unit of length 15
Avogadro constant 11, 52
Axial bonds (cyclohexane) 82, 149

Backward wave oscillator 18
Band centre 32, 33, 35
Bandwidth 19, 90
Base peak, in MS 106
Beer-Lambert Law 56, 57, 62
Bending, in IR 23, 27, 28, 30, 40–42, 44–46
Benzaldehyde, MS 119
Benzene 23, 42, 43, 51, 55, 60, 66, 73, 143
Benzene derivatives, IR 38, 42, 43, 45
  MS 107, 111, 114–116, 119
  NMR 73, 75, 76, 82–84, 96, 101
  UV 55–58, 60, 66
B_2H_6 89
Biological molecules,
  MS 121
  NMR 92,97
  X-ray 145, 146
Biomarker 120
Body-centred cubic cell 136–141
Boltzmann,

constant 12, 50, 70
distribution 12, 18, 70
Bond angles 1, 7, 23, 27, 131, 144
Bond length, dependence on vibrational
  energy level 1, 5, 7, 11, 13, 17–23, 24, 32,
    34, 36, 38, 48, 131
Bond lengths
  benzene, C-C 143, 149
  B_2H_6 89
  CO 10, 11, 35, 37, 38, 50, 51
  CO_2 149
  electron diffraction 147
  equilibrium 36–38
  F_2 20, 22
  Halogen bond 17
  HCl 49
  HCO^+ 14, 15
  OCS 50
  r_e, r_o, r_1 36–38
Bragg angle (θ) 134
Bragg equation 133, 133, 134, 147, 150
Bragg, William Henry 133
Bragg, William Lawrence 133
Bravais Lattice 139
Broadband 17–19, 24
Bromine compounds 74, 81, 82, 108, 118, 127
Buckminsterfullerene, C_{60}, 51, 149
Buta-1,3-diene, UV 60, 61
Butanone 92, 94, 95, 99
Butenone, UV 60

C_{60} 51, 149
^{13}C, NMR
  chemical shifts 93
  decoupling 92
  DEPT 95
  relation to structure 93–97
Carbon dioxide,
  bond length 149
  infrared 27–31
  Raman spectra 29, 30
  vibrational modes 23, 27
Carbon monoxide
  IR 30, 31, 34, 35, 50
  rotational spectrum 10–13
Carbonium ions 110, 111, 117
Carbon tetrachloride 149
Carbonyl complexes, IR 47, 48, 51
Carbonyl compounds
  IR 41–43, 45–48, 51
  MS 109, 111, 112–119
  NMR 71, 75, 80, 86, 87, 93, 94, 99,
    122–125, 127
  UV 54–58, 61, 62, 66
β-Carotene 62

Cartesian axes 16
CCD 133
Centre of inversion 29, 30
CF$_3$Cl 149
Chemical shifts
  ^{1}H NMR 72–75
  ^{13}C NMR 93, 94
Chiralamine X-ray 144
Chirped pulse18, 19
Chlorine compounds
  electron diffraction 149
  MS 105, 108, 113, 115
  NMR 74, 84, 85, 116, 127
Chlorobenzoic acid 115, 116
Chlorocyclohexane 151
Chlorine isotopes 105, 107, 108, 113, 116, 127
Cholesterol acetate, ^{13}C NMR 92, 93
Choline 99
Chromophores 57, 58, 60, 62, 65, 66
CI in mass spectrometry 105, 107
Cinnamic acid 83
CN Group 89
C–O, C=O bonds, IR 42, 43, 45, 46
  see Carbonyl compounds
Colour 52–55, 61, 62
Combination bands 38, 40
Combination
  differences, method of 36, 37
Conformations 82, 149, 151
Conjugation 57, 60, 61, 63
Constructive interference 134
Crick, Francis 144
Conjugation 57, 60, 61, 63
Continuous wave NMR 70
Copper
  lattice type 135, 136, 141
  X-rays from 131
COS OCS 50
COSY (NMR) 96, 97
Couplings (see splittings)
Cr$_2$O$_7^{2-}$ 65
Crystal systems 135
Cubic unit cells 135, 136, 138–143
Cumene 84, 100, 101
Cyclohexane 82, 149
  electron diffraction 148
  NMR 101
Cyclohexanone 59
Cyclooctatetraene 149

DEPT, in NMR 95
de Broglie118, 147
Decoupling in NMR 92, 94, 95
Degeneracy 12, 27, 47
Degrees of freedom, 4, 23
  electronic 2
  rotational 3, 6, 16, 23
  translational 3, 23
  vibrational 2, 3, 6, 26
Delocalisation 57, 61, 63, 73, 110
Deshielding, NMR 73, 74

Destructive interference 134, 135
Deuterium compounds 51, 70
Diborane 89
1,1-Dibromoethane 127
1,3-Dibromopropane 81
1,1-Dichloroethane 84, 85
Dichloromethane 127
1,3-Diiodopropane 94
Dienes, UV 59–61
Diethylamine 46, 113, 114
1,4 Dimethylbenzene, NMR 76
Diode array 54
Dispersion 54
Dissociation Enthalpy (energy) 26, 120
  equilibrium 26
  zero point 26
Dissociation limit 26
Distribution curve (ED) 148
DNA, X-rays 145

Effective mass 13, 14, 24, 25, 35, 39, 41, 51
EI mass spectrometry 103, 106
Electromagnetic spectrum 1, 2, 4, 5, 7, 8, 10, 19, 24, 31
Electron density maps 143, 145
Electron diffraction
  experiment 147
  examples 149
Electronic absorption
  examples 54, 61–66
  experiment 53
  indicators 56, 64
  keto-enol tautomerism, 63, 66, 86
  λ, ε structure 60
Electron limpact (EI) 103
Emission 2, 3, 13, 18, 19, 22, 31, 53, 70
Energy levels and changes,
  electronic 2, 4
  nuclear magnetic 3, 4
  rotational 4–10, 12, 16, 20, 31, 33
  vibrational 4–6, 11, 12, 24–29, 31, 33
  vibration-rotation 31–33, 37
Enols 63, 66, 86
Enzymes 66, 131
Equatorial bonds (cyclohexane) 149
Equilibrium bond length 36–38
ESI mass spectrometry 105, 121
Esters
  IR 42, 43, 46,
  NMR 75, 92, 93, 99, 100
  MS 112
  UV 59
Ethanal
  NMR 72, 77, 79
  UV 59
Ethanoic acid
  IR 45
  MS 107
Ethanol
  IR 39, 43
  MS 107

NMR 71, 91
UV 58
Ethers
  MS 113
  NMR 71, 75, 77, 78, 81, 113
Ethylamine, MS 117
Ethyl ethanoate, IR 46
Ethyl group
  MS 106, 109, 111, 113, 117, 118
  NMR 77, 78, 80, 99
Excited states 3
Extinction coefficient 52, 56

Face-centred cubic cell 136, 139, 140, 142, 143
Far infrared 7, 10, 11, 50
Fast-atom bombardment (FAB), 104, 121
Fermi resonance 30
Finger-print, IR 40
Flame test 2
Fluorescence 22, 53
Fluorine
  NMR (^{19}F) 88, 90, 91, 101
  Rotational Raman (F$_2$) 20, 21, 22
Force constant (of bond) 24, 25, 35, 39, 41, 51
Fourier transform (FT)
  microwave spectroscopy, 18
  IR 23
  NMR 31, 89–97
Fragmentation in mass spectra
  examples 109–119
  rearrangements 113
  simple fragmentation 109, 113
  peptides 121–126
Franklin, Rosalind 145
Fraunhöfer lines 2
Free Induction Decay (FID), 18, 90
Free rotor 7, 24
Frequency domain 18
Frequency of radiation 2–4
Fuel analysis 30, 31
Fundamental
  absorption 26, 29, 30, 38, 40
  transition 26, 28, 32
  wavenumber 26
  frequency 51

Gas analysis 30, 31, 119, 120
Gas chromatography 120, 121
GC/MS 121
GPC/MS 121

Haemoglobin, X-rays 145, 146
Halogen bonding 17
Harmonic oscillator 24, 25, 35
Hassell, Odd 149
HCO$^+$ microwave 14, 15, 17, 50
Hexane, IR 44
High resolution
  MS 105, 107, 119
  NMR 76
  Spectroscopy 5

Hodgkin, Dorothy 145
Hooke's Law 24
Hydrogen bonding
    IR 42, 43, 45
    NMR 85, 86, 88
Hydroxyl group 40, 42, 43, 45, 83, 85–88

Immunoassay 66
Indexing (X-rays) 139
Induced local fields, NMR 72, 73
Inductive effect, NMR 74, 81, 85
Infrared
    active 27–30, 38, 48
    inactive 27–30
    spectroscopy 24–28, 30–49
Infrared spectra
    diatomic molecules 24–26, 30, 34
    metal complexes 47
    polyatomic molecules 38–48
    triatomic molecules 27, 28, 30, 31, 34
Inelastic scattering 7, 8, 18–20, 28, 31
Integration trace, NMR 71, 72
Intensities
    IR 26, 27, 40, 47
    NMR 71, 89, 91–93
    Raman 20
    rotational spectroscopy 12, 20
    UV 52, 54, 57
Intermolecular interactions 6, 64, 86
Internal coordinates 23, 25, 38
Interstellar molecules 13, 14
Ionic crystals, X-rays 130, 136, 138, 140
Ionic radii 141, 151
Ionisation energy 120
Ionisation source 103, 104, 119
Iron carbonyl, $Fe_2(CO)_9$ 51
Isomers, of alkenes 83
Isopropylbenzene 85, 101
Isosbestic point 64
Isotope labelling, MS 120
Isotopes, in mass spectra
    bromine compounds 108, 116
    chlorine compounds 108, 113, 116
    M, M+1, M+2 peaks 107, 108, 113, 116
    neon isotopes 105
    sulfur compounds 108
Isotopic atomic masses 105
Isotopic substitution, microwave 14, 15, 17, 50
Isotopologue 50

J, rotational quantum number 8, 11, 12, 20
J_{max} 12, 13
J_{HH} (NMR) 79, 82, 83, 96, 97

K_α (X-ray) 131
Ketones: see propanone, etc
Keto-enol tautomerism 63, 66, 86
Klystron 7, 18

Laser desorption 105, 121
Lasers

Argon 22
HeNe 22
Nd:YAG 22
    in electron diffraction 150
Lattices in solids 135, 143
Lewis, G. N. 3
Lithium Chloride, X-ray 141, 151
Low energy electron diffraction (LEED) 150

McLafferty rearrangements 112, 115, 117
Magic angle spinning 91
Magnetic moments 69, 77, 90, 92
Magnetic Resonance Imaging (MRI) 98
Magnetic Resonance Spectroscopy (MRS) 98
MALDI mass spectrometer 104, 105, 107
Masses, atomic and molecular 105–107
Mass spectra
    analysis 107
    examples 109, 113, 117
    fragmentations 109, 113
    worked examples 127
Mass spectrometer
    double-focussing 104
    single-focussing 103
Mass spectrometry, Applications
    chromatograpy 120
    developments 119
    isotope-labelling 120
    molecules 106
    non-volatile compounds 121
    peptides 121–126
Maxwell-Boltzmann distribution 12, 70, 71, 144
Metabolites 99
Methylbenzene 45, 110, 111
Methyl butanoate, MS 112
Methylethyl benzene, NMR 85, 101
Methyl-4-bromobenzene, NMR 82
Methyl methacrylate 51 (answer to exercise 2.12)
Methylpentan-2-one, MS 115
Methyl red, UV-vis 56, 64, 67
Microwave radiation 4–10, 18, 19
Miller Indices 137
Millimetre-wave radiation 5, 7, 9, 10, 12–14, 18, 31
Millimetre-wave spectroscopy 7, 9, 14, 18, 31
MnO_4^- 65
Molar decadic absorptivity 52, 56, 57, 64, 66
Molar extinction coefficient 52, 56, 57, 64, 66
Molecular energy levels 53
Molecular formulae 107
Molecular ions, in MS
    determination of mass 103, 105
    ethanol 106, 109
    halogen-containing 107, 108, 113, 116, 118
    nitrogen-containing 109, 114, 116, 117
    sulfur-containing 108
Molecular orbitals 57
Molybdenum, X-rays 131, 140, 151
Molybdeum, organometallic complex 143

Moment of inertia 8–10, 13–16, 23, 32, 37
Monochromatic radiation 18, 22, 130, 131
MRI 98
MRS 98
MS/MS 121
Mutual exclusion principle 29, 30, 48

Neon, isotopes 103–105
Neutron diffraction 147
N-H bonds, IR 40, 43, 46, 48, 49
Nickel, dimethylglyoxime 65
Nitrites 93
Nitroalkanes, UV 58, 84
Nitrobenzene, UV 58
Nitrogen compounds
    IR 42, 46, 47
    MS 109, 113, 114, 117, 121, 122, 126, 128
    NMR 74, 75
    UV 60
    see also Amines, amides, peptides
Nitromethane, UV 60
1-Nitropropane, NMR 85
NMR
    ^1H chemical shifts 75
    ^{13}C chemical shifts 93
    continuous-wave 70
    decoupling 92–94, 95
    MRI 98
    MRS 98
    organic molecules 74
    pulsed spectrometer 89, 95
    spin-spin splittings 76
    two dimensional 95
Normal modes (vibration) 23–25, 27, 38
Nucleic acids, X-ray 145
Nuclear overhauser effect (NOE) 97
Nylon-6.6 47

OCS 50
Optical absorption 52
Optical density 54
Overtones 25–30, 38, 40
Oxygenation 98
Oxygen isotopes, in MS 106, 107

Pascal's Triangle/Pyramid 78, 79
Penicillin 146
Pentan-2, 4-dione,
    NMR 86
    UV 63
Pent-I-ene 60
Peptides,
    IR 46
    MS 121–126, 127 (Exercise 5.4)
Perspex 51
Perutz, Max 146
Phenylmethyl group 110
Phosphorescence 53
Phosphorus P_4 148
Phosphorus trifluoride 110 (Exercise 4.6)
Planck, Max 2

Planck's constant 3, 8, 9
Planes in crystals 137–143
Polarizability 19, 28, 30
Polymers,
    DNA 145
    nylon-6, IR 46, 47
    perspex, IR 51 (Exercise 2.12)
    proteins and peptides 46, 121–126, 128,
        131, 144–146
Population of energy levels 12, 18, 20
Positive ions, focusing 102–104
Potassium chloride, X-ray 140
Potential energy 24–26, 103
Primitive cubic cell 136
Propanal 59
Propanol 107
Propanone
    IR 39, 41
    UV 54–58, 60
Propanal 59
Propyl groups, NMR 65
Prostate 99
Proteins 121, 125, 131, 144–146
P,R branches 32, 34, 35
PSA test 66
Pulsed NMR 89–97
Pyruvrate 99

Quadrupole, MS 104
Quanta
    of radiation, 52
    energy 3, 9
    photons 3, 19
Quantitative analysis
    MS 120, 121
    NMR 63, 86
    UV 56, 57, 62–66
Quantum mechanics 1, 21
Quantum numbers 8, 25, 53
Quantum theory, energy levels 1, 2, 8

RADAR 7, 18
Radial distribution function 149, 151
Radiation, wave nature 2, 3
Radio astronomy 5, 13
Radiowaves 4, 5
Raman
    active 28, 29, 49
    Chandrasekhara Venkata 19
    effect 19, 20
    inactive 28, 30
    scattering 8, 19, 22, 28, 29
    spectroscopy 5, 7, 19–23, 28–30
Rayleigh scattering 19–21, 28, 29
Rearrangements (in MS) 111,–113, 115
Reciprocal moment of inertia (see rotational
        constant) 9
Reduced mass 8, 10, 11, 35
Relative atomic mass 105–107
Relative molecular mass 105–107

Residual mass (amino acids) 121, 122
Restricted rotation 87
Rigid rotor model 7, 8, 10, 13
Ring current (NMR) 73
Ring substitution pattern 42, 116
Rotational
    constant 7, 9–16, 21, 22, 32, 34–37
    energy levels 5, 7, 9–12, 16, 19, 32, 53
    quantum number (J) 8, 11
Rotational Raman spectroscopy 7, 19–23
Rotational spectroscopy
    of diatomic molecules 6–13
    of linear triatomic molecules 13–15
    of non-linear polyatomic molecules 16, 17
Rotational term expression 9

Sanger, Frederick 125, 145
Scattering (diffraction) 130
Scattering, elastic and inelastic 3, 7, 8, 18, 19,
        20, 22, 23, 28, 29, 31
Selection rules 9, 10, 13, 25, 26, 32, 49
Shielding (NMR) 72–74, 76
Sodium chloride
    analysis of structure 142
    single crystal X-ray 133
    X-ray powder 132
Sodium flame, colour 2
Sodium ion, ionic radius 151 (Exercise 6.3)
Spherical rotor 16
Splittings (NMR),
    measurement 79
    origin 76
    worked examples 80
Stability of positive ions (MS) 109–111
Stokes scattering 19–22, 28, 29
Stretching (in IR spectra) 23, 24, 26–28, 30,
        39–48
Styrene (NMR) 83, 96, 97
Sulfur compounds, MS 108
Sulfur, isotopes of, MS 108
Supersonic free jet 18
Symmetric rotor 16
Symmetric top, 16, 17
    oblate 16
    prolate 16
Symmetry 1, 28, 29, 38
Systematic absence (X-rays) 139

Tables of
    chemical shifts (NMR) 75, 93
    IR absorptions 42
    residual masses (amino acids) 122
    UV absorptions 60
Tandem MS/MS 121
Tautomerism, keto-enol 63, 86
Testosterone 62
Tetramethylsilane, TMS 73
Thomson, J.J. 102
TOF, mass spectrometer 104
Transmittance

IR 11
UV 54
Trichloromethane 16, 74
Tryptophan 51, 124
Tumours 99
Two-dimensional (2D) NMR 95

Ultraviolet (UV)
    light 19
    region 45, 52–54, 57, 58
UV/vis
    spectroscopy 53–57
    characteristic absorptions 60
Unified atomic mass unit, 15
Unit cells, types and dimensions 135

Vibration-rotation,
    energy levels 31
    interaction constant 37
    spectroscopy 30–37
Vibrational,
    bands 29–31
    degrees of freedom 3, 6
    energy levels 4, 5, 11, 12, 24–26, 28, 31
    frequencies 5, 23–25
    normal modes 23, 24, 27, 28
    quantum number 25, 26
    spectroscopy 4, 23–48
    term values 25
    wavenumber 24
    zero point level 11, 12, 25, 26, 32, 34–38
Visible
    light 2, 4, 22
    photons, scattering 7, 8, 19, 31
    region 4, 26, 52–54, 57, 58
    UV-vis spectroscopy 53–67
Vitamin B_{12} 146
von Laue, Max 130

Water, IR 23
Watson, James 145
Wave-nature of radiation 2, 3
Wavelength of maximum absorption 55–57
Wavenumber, in IR 4, 9
Wollaston, William 2

X-ogen 14
X-ray diffraction,
    Bragg equation 133
    electron-density maps 143, 146
    examples of photographs 132, 133, 146
    generation 131
    apparatus 131
    ionic radii 141, 151 (Exercise 6.4)
    single-crystal studies 133
    structure (ionic) 138
    structure (molecular) 143

Zero point level:
    see Vibrational